Autonomous Real-Time Testing

Testing Artificial Intelligence
and Other Complex Systems

Thomas Michael Fehlmann

Logos Verlag Berlin

λογος

Bibliographic information published by Die Deutsche Bibliothek

Die Deutsche Bibliothek lists this publication in the Deutsche Nationalbibliografie; detailed bibliographic data is available in the Internet at http://dnb.ddb.de.

ISBN 978-3-8325-5038-7

Logos Verlag Berlin GmbH
Comeniushof, Gubener Str. 47,
10243 Berlin

Tel.:   +49 (0)30 / 42 85 10 90
Fax: +49 (0)30 / 42 85 10 92
http://www.logos-verlag.de

WHY THIS BOOK?

Readers who worked through "Managing Complexity" (Fehlmann, 2016) – the previous book that appeared end of 2016 – may already have been waiting for its continuation, addressing today's deadlock around digitalization. Combining *Artificial Intelligence* with autonomous vehicles and the Internet of Things creates new potential products. While technology is here for building autonomous vehicles that are much safer than human-driven cars, saving lives and eliminating traffic jams at once, society is all but ready.

Autonomous cars will not hit the roads within foreseeable time because of liability concerns. Who is responsible for incidents that they cause? The supplier of software, or hardware, or the user sitting in the passenger seat, or always the other, mostly unlucky, human involved?

And what if two autonomous cars crash into another?

Sure, nothing can avoid future incidents even with the best-equipped cars of the universe. However, precautions can be taken, and the most obvious precaution is testing of the software that affects behavior of the autonomous system. Since one of the most intriguing safety issues is with privacy violation, if data is stolen, or even malicious forces take over control, privacy protection is among the most important safety requirements for software-intense products. This book explains also how to test privacy protection and assess safety risks, but its main purpose is to explain how automated autonomous testing works, in real-time, at the users' premises.

Because, today's software testing is way behind the age of digitalization. No metrics exist for test intensity that can compare different cars and manufacturers. Today's software tests cover code, but it is unclear whether the full functionality needed to control autonomous systems is tested at all. While software changes continuously, with *Continuous Integration/Continuous Delivery*, tests executed at release quality gates reflect the original state of delivery, in isolated environments. Autonomous systems affect the real world, unfortunately. Thus, *Continuous Testing* is requested.

This book is not an experience report but creates a vision. It proposes a way for implementing continuous and autonomous real-time testing for software-intense systems.

All examples shown are implemented in Excel in Microsoft Office 365 and freely available to readers of this book, including technical information. It suffices to send e-Mail to info@e-p-o.com with some evidence of purchase and a valid e-Mail address. Access is personal, encrypted and protected. This is necessary since the examples contain open VBA code that otherwise can be compromised.

ACKNOWLEDGEMENT

Important contributions originate from discussions and workshops with colleagues from the Software Metrics and the Quality Function Deployment communities; especially, but not limited to, Silvan Fehlmann, Luigi Buglione, Eberhard Kranich and the German QFD Institute.

Special thanks to Eberhard Kranich who lectured the book.

TABLE OF CONTENTS

CHAPTER 1: WHY AUTONOMOUS REAL-TIME TESTING?

Autonomous Real-time Testing sounds somewhat like one of the many hypes that currently come with digitalization. The strange effect originates from the term "Testing" – something that sounds nowadays somewhat outdated. Who is interested in Testing? Agile Enterprise, Agile Management, DevOps, Industry 4.0, Disruptive Transformation are stirring more interest.

However, most products today are software intense. Such products, as any product, might fail, and if such failure causes damage or loss of goods or life, liability questions arise. The Internet of Things (IoT), Advanced Driving Assistance Systems (ADAS), Autonomous Drones for goods delivery or building industry, all are under the thread of failure caused by software and – consequently – by liability issues.

This book does not address hardware failure, or failure by mechanical design or construction. The focus is on failure by software faults, and what else can we do than software testing against failure? When should we do such testing? At the end of software development? When does development stop with DevOps? Should we probably add Continuous Testing to Continuous Integration and Continuous Delivery (CI/CD/CT)?

1-1 INTRODUCTION

The first topic to address is where did our famous software projects go that where always too late, with cost overruns, and tests left to the first customers?

With DevOps, there are no more projects. DevOps is a paradigm for product management by continuous software integration and software delivery. The end of development is not before end of product life. While there is still product design and software architectural thinking, new software is created by integrating open source software with its own coding. Software testing is difficult, and since testing refers to code, only the part written in-house undergoes testing. Each build is fully unit tested, supported by test automation tools to ensure that code meets expectations. Test coverage refers to code; necessarily only to that part of software written in-house. Technical debt is the common metrics for released code that requires rework, be it for maintainability or extensibility. The metric describes the amount of effort needed to remove weaknesses and is typically calculated by the code repository system, e.g., SonarCube

(SonarSource S.A, Switzerland - Open Source, 2017). The aim is to totally avoid technical debt.

Also, unit testing of code written by the team is normally done daily; and no code can be checked-in that does not pass unit tests. Best practices ask for unit tests written before any code. This is the *Test-Driven Development* (TDD) approach made famous by the Poppendiecks (Poppendieck & Poppendieck, 2007).

Final release testing for the product is done where compliance issues exist that need being verified. Otherwise, system testing is usually done together with customers; most often, but not always, these customers are aware of acting as a "beta-tester". System testing cannot be performed by the code testing tools used for unit testing where code is not available. Test coverage remains guesswork even after intense and effectively monitored beta tests with users. No test metrics exist beside unclear indications like the number of bugs detected and recorded in some issue management tool. Since bugs can neither be identified and located in code, nor separated from each other, the number of bugs recorded is a useless count. Whether two bug entries refer to the same defect or not remains open.

While for game software or entertainment, even office publishing, such a situation is acceptable, it is clearly not for products based on software that carries liability. Home banking software without defect density measurements is risky for banks. If office software becomes a tool for team communication in enterprises, liability for its correct functioning carries a significantly higher risk for the software supplier than document publishing and spreadsheet calculating software. Software that controls the *Internet of Things* (IoT), or even more for *Advanced Driving Assistance Systems* (ADAS) in cars or autonomous vehicles might have disastrous effects if not working safely and correctly. All these examples do not rely on code written by some single development team. They are rather a patchwork of functionality delivered with the "thing" or system component; whether code is available, is uncertain. The need for testing is apparent; the need for test metrics that characterize the amount of testing and current defect density is obvious. However, while casual testing might be done somehow by suppliers and users of such software; metrics are not available and not agreed. Without metrics, such casual testing is near to useless.

Yet another problem lies with software borrowed as services from the cloud. For instance, communication software might be vulnerable to data theft; social media and team communication might be subject to unauthorized big data analysis violating privacy rights; assumptions for cloud service security might turn out to be overly optimistic without testing and test metrics. Consequently, autonomous vehicles might take the wrong route, or keep routes taken not private. This is the small side of the problem; safety risks by untested software-intense systems constitute the big end.

While privacy and safety risks are not the whole story related to digitalization, these two topics embrace the most urgent need for systems testing in software-intense products.

1-2 WHAT IS SOFTWARE TESTING?

Software Testing means the process of defining *Test Stories* (or *Test Scenarios*), each containing *Test Cases*, and execute them with the aim of detecting unexpected behavior. A *Test Case* is a structure consisting of *Test Data* $x_1, x_2, \ldots, x_n$ and a *Test Response y*, where each test data item x_i as well as the test response is an *Assertion*. The assertion describes the state of the program under execution. Formally, a test case is expressed by the following *Arrow Term*:

$$\{x_1, x_2, \ldots, x_n\} \rightarrow y \tag{1-1}$$

For the origins of arrow terms see Engeler (Engeler, 1981). For a more recent application, how arrow terms define a neural algebra on "how does the brain think?", see again Engeler (Engeler, 2019). In our case, the assertions describe the status of the software-intense system under test. A simple assertion describes the value, or value range, of a software variable; it can also describe a certain status of the system, such as listening to some device, waiting for confirmation or executing a database search, or simply identify the starting point for some test case. Related to Six Sigma, the left-hard finite set of an arrow term is referenced as *Controls*, the right-hand singleton is the *Response*.

Assertions use the basic numerical operations between variables and constants such as equality, greater than, or inequality. It is not necessary to combine assertions using logical operations AND, OR, and NOT. The test data sequence within the left-hand side acts as an AND. Instead of an OR, use two arrow terms. NOT is more complicated to substitute by arrow terms: sometimes, negation is immediately available as with equality, sometimes, negated assertions split into two or more positive assertions. The test response y is not necessarily unique; several assertions might become true under identical test data assertions $x_1, x_2, \ldots, x_n$, for instance depending where the system under test is investigated for the test result.

A test case passes if we can run the software with valid test data assertions and the assertion y for the test response is valid in the system under test. A test story passes if all its test cases pass. *Real-time Testing* is the process of testing real-time systems and its software, see Ebner (Ebner, 2004).

Assertions may include stronger assertions. For instance, the assertion $a = 20$ is more restrictive than $a \leq 20$. Test cases always contain weakest assertions; thus, inequalities or range specification rather than sample numbers.

1-2.1 A Standard for Representing Assertions about Tests

Since test cases are possibly something that shall be exchanged between different systems, even from different manufacturers, standardization is needed. If software from different suppliers shall cooperate, standards must be agreed and implemented that allow communication and cooperation. In the IoT and automotive area, such standards exist. For real-time testing, with focus on communication, an international standard for specifying test cases exists: *Testing and Test Control Notation* (ETSI European Telecoms Standards Institute, 2018), now in its version 3 (TTCN-3). According Ebner (Ebner, 2004), the test notation is useful for automatically generating test cases from UML sequence diagrams, covering the base system. In our context, TTCN-3 is suitable for stating assertions. However, TTCN-3 is much more than simply a framework for stating test assertions such as fixing test data and test responses. It also contains the necessary instructions for test instantiation and test automation.

Thus, using TTCN-3 for test assertions, software tests can be described by a standard that is independent from the programming environment and from the supplier. Tests can be interchanged between different actors related to software testing.

1-2.2 A Representation for the World of Tests

However, software is dynamic. Trying to model software by static assertions is missing the dynamic behavior of a system. For this reason, we extend our definition of a *Test Case* to include not only basic assertions but recursively other test cases as well.

Let $\mathcal{L}$ be the set of all assertions over a given domain. Examples include statements about customer's needs, solution characteristics, methods used, etc. These statements are assertions about the business domain we are going to model. A sample language $\mathcal{L}$ is TTCN-3. However, since this book is written for humans, not robots, we will use natural language, not TTCN-3, knowing that our verbal assertions need being converted in machine language before being executed.

Denote by $\mathcal{G}(\mathcal{L})$ the power set containing all *Arrow Terms* of the form (1-1). The left-hand side is a finite set of arrow terms and the right-hand side is a single arrow term. This definition is recursive; thus, it is necessary to establish a base definition saying that every assertion itself is considered an arrow term. The arrows of the arrow terms are distinct from the logical imply that some authors also denote by an arrow. The arrows denote cause-effect, not a logical implication.

The formal, recursive, definition, in set-theoretical language, is given in equation (1-2):

$$\mathcal{G}_0(\mathcal{L}) = \mathcal{L}$$
$$\mathcal{G}_{n+1}(\mathcal{L}) = \mathcal{G}_n(\mathcal{L}) \cup \left\{ \{a_1, \dots, a_m\} \to b \,\middle|\, a_1, \dots a_m, b \in \mathcal{G}_n(\mathcal{L}), m = 0,1,2,3 \dots \right\}$$

$$(1\text{-}2)$$

$\mathcal{G}(\mathcal{L})$ is the set of all (finite and infinite) subsets of the union of all $\mathcal{G}_n(\mathcal{L})$:

$$\mathcal{G}(\mathcal{L}) = \bigcup_{n \in \mathbb{N}} \mathcal{G}_n(\mathcal{L}) \tag{1-3}$$

The elements of $\mathcal{G}_n(\mathcal{L})$ are arrow terms of level n. Terms of level 0 are *Assertions*, terms of level 1 *Test Cases*. Sets of test cases are called *Rule Set* (Fehlmann, 2016). In general, a rule set is a finite set of arrow terms. We call infinite rule sets a *Knowledge Base*. Hence, knowledge is a potentially unlimited set of rules about assertions about test cases. This definition is recursive, too.

A *Test Story* is a finite rule set and element of $\mathcal{G}_n(\mathcal{L})$ that consists of level 1 terms only. A test story comes with additional information relating to its business domain.

1-2.3   Combining Tests

Let M, N be two rule sets, consisting of test cases. N is a set of test cases consisting of arrow terms of the form $b_i = (\{x_1, x_2, \ldots, x_n\} \rightarrow y)_i$. Then application of M to N is defined by

$$M \bullet N = \{c \mid \exists \{b_1, b_2, \ldots, b_m\} \rightarrow c \in M; \ b_i \in N\} \tag{1-4}$$

In other words, if all b_i executed in N with pass, the test story M can be applied to a rule set N as a set of test cases. This represents the selection operation that chooses those rules $\{b_1, b_2, \ldots, b_m\} \rightarrow c$ from test story M that are applicable to the rule set N. Combining tests is a strong means to extend test stories up to the limit as needed.

Combinatory Algebra (Engeler, 1995) is the mathematical theory of choice for automatically extending test cases from a simpler, restricted system, to test stories that fully cover a larger, expanded system. The extension works only if software testing is not only automated but also measured. Metrics must be independent from current implementation and from actual system boundaries.

The theory of *Combinatory Logic* postulates the existence of *Combinatory Algebras* whose computational power is Turing-complete: i.e., all programs that are executable by computers can be modeled. This guarantees the best achievable test coverage. With combinatory algebra, test cases extend from real-time tests, covering a base system, to the actual, expanded system.

The definition (1-4) looks somewhat counter-intuitive. To apply one test case to another, it is required that the result of application contains all the full test cases providing the response sought.

A more intuitive approach would only require arrow terms providing such a response meeting the required controls. The existential quantifier would then guarantee that there is a test case yielding such response. When accepting the axiom of choice in its traditional form, that does not look like a problem. However, it is left to the interested reader to prove that this seemingly more intuitive approach would immediately lead to a contradiction to Turing's halting problem (Turing, 1937).

Since we are computer scientists and not traditional mathematicians, we require the intuitionistic, or constructive, variant of the axiom of choice. The existential quantifier requires not only the existence as such, but construction instructions for the existence of arrow terms. It means for test cases, that it is not enough to know the existence of tests, but you need to know how to construct them. This is the reason why our formal system for automated testing requires at least level 1 – arrow terms for applying one test set to another – and this is possibly also the reason why test automation has proven to be so hard.

And for those who consider such reasoning too theoretical, let's provide a rather practical argument: programmers who want to set up test concatenation $M \bullet N$ for automatic testing, need access to the test cases in N that provide the responses needed for M. Otherwise combining tests is unsafe or cannot be automated. Thus, with the combinatory algebra of arrow terms, mathematical logic meets intuitionism and practical programming.

1-2.4   ARROW TERM NOTATION

To avoid the many set-theoretical parenthesis, the following notations are applied:

- a_i for a finite set of arrow terms, i denoting some finite indexing function for arrow terms.
- a_1 for a singleton set of arrow terms; i.e. $a_1 = \{a\}$ where a is an arrow term.
- $\emptyset$ for the empty set, such as in the arrow term $\emptyset \to a$.
- (a) for an (potentially) infinite set of arrow terms, where a is an arrow term.

The indexing function cascades, thus $a_{i,j}$ denotes the union of a finite number of m arrow term sets

$$a_{i,j} = a_{i,1} \cup a_{i,2} \cup ... \cup a_{i,j} \cup ... \cup a_{i,m} \tag{1-5}$$

With these writing conventions, $(x_i \to y)_j$ denotes a rule set, i.e., a finite set of arrow terms having at least one arrow. Thus, they are level 1 or higher.

With this notation, the application rule for M and N now reads

$$M \bullet N = \{c | \exists b_i \to c \in M;\ b_i \subset N\} \tag{1-6}$$

Or, in an abbreviated notation:

$$M \bullet N = (b_i \to c) \bullet (b_i) \qquad (1\text{-}7)$$

Arrow terms are not only useful for representing test cases. *Quality Function Deployment* (QFD) is a well-known method for customer-oriented product development (ISO 16355-1:2015, 2015). Each element $x_i \to y$ of $(x_i \to y)_j$ denotes one Ishikawa diagram (Akao, 1990), which is a cause/effect constituent of a QFD deployment and stands at the origins of QFD in Japan. The matrix $(x_i \to y)_j$ represents the QFD deployment. This matrix obviously is a rule set within $\mathcal{G}(\mathcal{L})$. The union of all possible QFD matrices is infinite and therefore a knowledge base in $\mathcal{G}(\mathcal{L})$.

Six Sigma Transfer Functions are constructively defined functions $\boldsymbol{A}$ used in the form $\boldsymbol{y} = \boldsymbol{Ax}$, where $\boldsymbol{y}$ is the observable response, and $\boldsymbol{x}$ is the vector of unknown causes. For a short primer on transfer functions see section 2-3. Each set of arrow terms represents a constructively defined Six Sigma Transfer Function. This was originally described by Ishikawa (Ishikawa, 1990).

The *Ishikawa Diagram* (Ishikawa, 1990) describes the cause-effect relations between topics and are considered the initial form of QFD matrices, and thus of linear transfer functions. Converting a series of Ishikawa diagrams into a transfer function is straightforward, see Figure 1-1 below. Rules correspond to the cause/effect correlations.

Figure 1-1. Representing Transfer Functions as Rule Sets

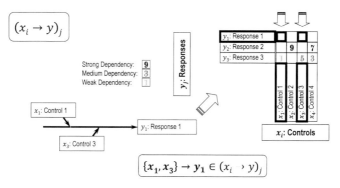

Each element $x_i \to y$ of $(x_i \to y)_j$ denotes one Ishikawa diagram (Akao, 1990), which is a cause/effect constituent of a transfer function. The matrix $(x_i \to y)_j$ represents the full transfer function. Transfer functions obviously are rule sets within $\mathcal{G}(\mathcal{L})$. The set of all linear transfer functions is infinite and therefore a knowledge base in $\mathcal{G}(\mathcal{L})$.

Other elements of $\mathcal{G}(\mathcal{L})$ do not resemble linear transfer functions, such as

$$\big((x_i \to y)_j \to z\big)_k \qquad (1\text{-}8)$$

This is a finite set of arrow terms whose left hands consist of finite rule sets. Another such example is $x_i \rightarrow (y_j \rightarrow z)$. This is a cascade of rules. The association for arrow terms is to the right:

$$x_i \rightarrow y_j \rightarrow z = x_i \rightarrow (y_j \rightarrow z) \tag{1-9}$$

1-2.5 TEST AUTOMATION

Tools used for implementing such an approach are test stories and test cases that use a formal language to be machine-readable. The language is implemented as *Arrow Terms*, see Engeler (Engeler, 1995), a model of combinatory algebra describing the general Six Sigma approach, listing controls for observable responses of a system. Responses can be multi-dimensional, resulting in a *Response Profile* that is measurable and thus can be compared to the expected response.

For the mathematical structures of *Six Sigma Transfer Functions*, see Fehlmann & Kranich (Fehlmann & Kranich, 2011) and Fehlmann (Fehlmann, 2016)\d. Transfer functions are used to uncover search response controls, for instance in Google search requests, or technical solution that meet customer's needs. In testing, transfer functions indicate whether the goal of testing is achieved. The degree of achievement is called *Test Coverage*. Test coverage can control automated generation of meaningful test cases in a chosen context. Automatically generated test cases are selected only if they contribute to the testing goal.

1-2.6 EXECUTING TESTS

Since arrow terms define test data up to an assertion, in ordered domains such as numbers, test data may be defined only up to some range. Thus, when executing the test, there is a choice which data exactly to select. If the range is limited, it is straightforward to select the limit, or possibly to explore the numerical precision of the limit. Thus, the code implementing the test case may need more than one execution when running the test case. However, we count the test case only once even if its execution requires multiple runs.

1-3 REPRESENTING UNLIMITED KNOWLEDGE

Rule sets represent things that organizes themselves such as cars that drive automatically, flying drones that find the way to its target, smart homes that save energy. These things typically acquire knowledge while they are in operations. Predicting their behavior is ultimately impossible without representing the knowledge acquisition during operations.

Interestingly, agile software development works the same way: exact specifications are unknown at the beginning. While software is developed together with the stakeholders, more and more the ultimate result becomes apparent. Combinatory Logic thus looks interesting as a framework for better understanding and modeling agile software development.

1-3.1 PARALLEL COMPUTING

Rule sets are of unlimited size but well structured. Moreover, if the base set represent transfer functions, they carry associated metrics, namely the *Convergence Gap*. Successful software testing relies on measurable cause-effect relationship.

There are various measures that can be applied: functional size, security, safety, cost, non-functional metrics such as ease-of-use. The IoT consists of things made intelligent by software, connected by software, and acting autonomously by software. This is called an *IoT Concert*. Organizing an IoT concert is called *IoT Concertation*. The IoT concert is a valid paradigm for today's software-intense systems. Its main characteristics are it always grows, never being finished. Based on software metrics, two arrow terms describing software can be compared with respect to size, to defect density and compared with respect to behavior towards the same goal.

Behavior of an IoT concert changes when the environment changes – adding or removing things might change, or even create totally new behavior. Totally unexpected situations might emerge on streets driven by autonomous cars. The rule set is not completely known at any time; however, directed by metrics, a sufficiently good approximation can be built just when needed.

Implementing a rule set is by constructing an automaton that eventually produces all its elements. The arrow term notation (1-1) describes the algorithm needed for the automaton. The automaton produces arrow terms in parallel and in any order, without knowing much from each other. To make them useful, the automaton needs guidance through metrics-based heuristics.

1-3.2 THE RULE SET RADIUS AND VARIANCE

The trick is combining the strict and well-known structure of a rule set with the aim of the test. This requires understanding what the test should prove. The gap between aim and the test capabilities is expressed as the *Convergence Gap* of the test coverage matrix. The details are explained in section 2-3: *A Short Primer on Six Sigma Transfer Functions*.

The rule set can be constructed by an automaton that eventually produces each element after some time. The convergence gap directs that automaton to produce the arrow terms. It is possible to do this in predictable time (Fehlmann, 1981).

The arrow terms arise from asking the components of an IoT concert how they behave in some given circumstances. Asking the right question will do:

$$\{(a_i \rightarrow b)_j \,\big|\, \|b_j - \boldsymbol{\tau_y}\| < \varepsilon\} \tag{1-10}$$

where $\boldsymbol{\tau_y}$ is the *Goal Profile*, representing the target for the circumstances under investigation, and $\|...\|$ represents the *Euclidian Norm* for vectors. For instance, $\boldsymbol{\tau_y}$ might represent the condition that the autonomous car avoids crash. Then, equation (1-10) represents all crash-free conditions achievable by the autonomous car.

The controls to consider depend from the goal. It must be known what the goal of the behavior is: doing no harm or minimizing it for autonomous cars, minimizing energy consumption in intelligent homes, avoid crashing for flying drones. On the other hand, testing aims at finding fault conditions.

Rule sets consist of solution topics vectors. The convergence gap against the goal response vector can be computed, based on the achieved responses. Let $\Delta_1, \Delta_2, ..., \Delta_n$ be these differences.

$$Rule\ Set\ Radius = \text{Max}_{j=1..n}(\Delta_j)$$

$$Rule\ Set\ Variance = \sqrt{\frac{\sum_{j=1..n} \Delta_i^2}{n-1}} \tag{1-11}$$

Figure 1-2 demonstrates convergence gaps for three dimensions. Higher dimensions are more difficult to visualize but equally simple to calculate.

Figure 1-2. Small and Large Rule Set Radius and Variance for Three Dimensions

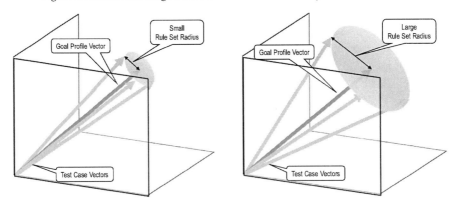

The *Rule Set Radius* is the maximum of all the convergence gaps in a rule set, thus acting as an envelope around them, serving as an indicator for total variations within a rule set.

The rule set radius has a strange similarity with the *Schurr Radius*, used for assessing consistency in AHP, see Schurr (Schurr, 2011) and Fehlmann (Fehlmann, 2016). What it means, is yet an open question. Maybe the rule set radius is an indication for the inconsistency of the test cases within a rule set?

1-4 AUTONOMOUS REAL-TIME TESTING

Testing becomes autonomous if test cases are no longer prepared a priori, but automatically generated while the system extends by connecting with new components or learning new things. This happens during normal operation. With the IoT, for instance, when adding some new IoT component. If a car meets another car of different making, or when different software releases meet, the system changes as well. *Autonomous Real-time Testing* (ART) means that new test cases are generated, and tests are executed, all in real-time, before allowing the new components to join, or to have impact. The time limit is needed for practical purposes. Systems supported by ART behave intelligent in the sense that they can anticipate the effects of actions even in previously unknown circumstances.

The base elements for autonomous testing are test stories, containing test cases, that cover the initial base system. A set of test metrics is needed for assessing the test intensity and defect density of the base system. According ISO/IEC 14143 (ISO/IEC 14143-1:2007, 2007), software metrics are independent from actual software implementations. Metrics must carry over from base tests to expanded automated tests covering the expanded system.

The main problem with testing the expanded system is how to generate new test stories that keep the focus on the relevant testing goals. Time and resources are limited. Sometimes, only a few seconds are available for generating and conducting real-time test runs. Measuring test intensity and predicting remaining defect density for the expanded system is necessary to understand the effect of actions taken by the expanded system. This compares the reliability of the expanded system with the original base system.

Another major problem is computational speed. Test execution includes searching suitable new test cases, executing them by asking the involved object, what they would do in such a case, and calculating results. The generation of test cases for the expanded system must deliver results in real-time. Autonomous real-time software testing is impossible without suitable, implementation-independent test metrics.

The rich structure of arrow terms on top of TTCN-3 is the ideal framework for autonomous testing. Test rules are potentially infinite sets of test cases that can be exploited for determining test cases suitable for enhancement. Autonomous testing always starts with a normal test; thus, with a finite subset of test rules belonging to a test story. Test cases can be added to such a test story, increasing test coverage and the capability to detect defects. Or, two groups of test cases can be combined based on equation (1-3) provided they belong to the same test story, i.e., they test the same business goal.

Setting up test rules for a software-intense system is now just the first step towards continuous testing. The test rules can be made permanently available to users for the entire life cycle of a product such that users can always verify that the product still behaves as initially convened. This is autonomous testing by users who are not testers. They can run the original test again and again. While this has some value already, it is not yet ART. The problem is that software is subject to the condition of the real world. In particular, the real world is changing over time quite a bit. Each software update that is downloaded has the capability to affect the behavior of the system. This is especially important for IoT concerts, where adding or removing a component can change the behavior, especially exposure to privacy intrusions or safety risks. But also, cars that talk to the smartphone are affected; and even more, systems that communicate with each other for instance to exchange traffic information. Since threats also change, new test cases are needed to detect new threats.

If new defects are detected, the result of the autonomous real-time test is shared with the software supplier for removal of the defect. Moreover, any such detection is shared with other users of the product. Testing is no longer an isolated activity of some group of testers that wait for a product to be ready for test; it becomes a community-based activity and closely linked to support and marketing of software-intense products.

1-4.1 TESTING SOFTWARE-INTENSE SYSTEMS

Obviously, the complexity that users can handle when performing tests is limited. The limit is closely related to the users' needs when using a product. While technical performance or other quality aspects are usually important when buying the product, during normal and intended use, other needs become dominant. Continuous availability of the functionality purchased is expected, and often the user perceives lack of basic functionality immediately. However, missing functionality related to communication and traceability is typically seen too late.

Defects that consumers affect but are hard to detect early enough include missing privacy protection, or safety issues. Both, privacy protection and safety risk assessments have become controversial and find today public attention. The fact that today all physical movements are traceable by the smartphones' build-in GPS, or the navigation instruments in cars, or soon by public transportation, seems only a weak violation of privacy compared to all the spying of what we look at, search for, read or talk to. It seems that if it does not touch apparently property, or money, we are inclined to weight the benefits of all these chatty assistants and free services higher than concern for privacy.

However, things change if one of these helpful service providers suddenly suffer a data leak, as happens from time to time indeed. Then it turns out to be quite hard to find out if own private data has been affected. Often, after a leak, it takes weeks if not months to find out what data was compromised.

ART in turn is something that can be immediately triggered in case some service provider experiences a hack. That would allow consumers to get immediately notified whether they need to take some action. Obviously, testing is not restricted to the local components of some system; this is sort of unit testing anyway. Testing a software-intense system today almost always involves testing services, typically located in the cloud. ART can check whether such services still behave as expected or start exhibiting strange behavior such as scanning the private device for passwords or opening new backdoors.

Similar it is for safety. Often, safe behavior is easily detectable and distinguishable from unsafe behavior. Continuous safe behavior even after software updates or changes in the related cloud services is testable and consumers are interested in such tests; although, full security testing includes hardware and is not addressed in this book.

Sure, privacy and safety are not the only test-worthy characteristics of software-intense systems, but they address the major concerns of most consumers. And indeed, having privacy tests performed after each major software update even for a smartphone would even now be a welcomed gadget. Or does anyone know for sure what privacy risks all those glossy games and racy apps entail?

1-4.2 CONSUMER METRICS

Testing is not for free, and when it does not add value to the product, it possibly should not be done. However, whether safety and security testing – among other testing activities – add value to a product, is out of question. Whatever adds value, can be used as a discriminator in the market.

It is therefore paramount that tested software becomes visible to the consumer, and the amount of testing becomes a metric communicated to consumers. Consumer might have a choice between an extensive, well-tested product and a less tested, but cheaper product. This works up to the point where liability issues force the product supplier to perform extensive testing just to stay in the market.

For all this IoT-kind of products with extensible software, especially for AI, it is inconceivable how suppliers should cope with liability if they do not have the possibility to do ART during their products' lifetime. The risk, that consumers add features or functions to an IoT concert causing harm to safety or privacy of the product is significant.

1-4.3 IMPACT ON SOCIETY

Autonomous Real-time Testing (ART) will make an impact on product liability issues of software-intense systems. Suppliers can reduce their liability risks when providing sufficiently good and actual test rules for their products during the full life cycle. Otherwise, owning or using an autonomous car – if they ever hit streets – will probably become quite costly, at least in Europe. Autonomous vehicles apart from closed motorways face many challenges and their advantages are limited; for instance, regarding an eventual safety advantage of autonomous vehicles, traffic in cities or villages becomes safer at less cost by reducing speed. Communication between vehicles is a big asset and could improve traffic flow even in urban areas; however, this means not necessarily autonomous driving.

In any case, autonomous vehicles, vehicles that rely on intercommunication and even driving assistance systems become socially more acceptable if they adapt their capabilities over time to changed environments and prove this to the responsible owner.

The adoption of the *Internet of Things* (IoT) is far below expectations not only because users wait for the faster and more performant 5G telecom network, but even more because the target users cannot assess and manage their privacy risk. Connecting an additional device to their existing IoT concert might result in an unnoticed privacy break. Only experts may give it a try.

Lessening the liability burden to suppliers of software-intense products clearly speeds digitalization up and make it more acceptable. Furthermore, since many of the new software-intense products use *Artificial Intelligence* (AI), such products change behavior during their life cycle and you cannot use AI in products without at least basic ART, at least for safety-critical issues. Deep learning is also accompanied with forgetfulness and even human neural disorder (van Gerven, 2017). Consequently, visual recognition systems need constant testing for making sure they keep their initial capabilities. Neural networks that have previously proven to be capable of successful learning suffer from strange effects (Szegedy, et al., 2014). Small alterations in images or video, even when invisible to the human eye, can strongly impact their capabilities.

1-5 OUTLOOK

From this introduction, many open questions arise. First, how shall tests be assessed such that they can be assessed for intensity and defect density? By measurements, possibly? Next, how to identify relevant tests in the huge test rules set generated by combinatory algebra? While having all tests at disposition in a structure helps, how to extract those tests that are relevant for privacy and safety – or any other goal?

Testing works only if the goal of testing is known. What suitable means exist to define goals of testing? What are the goals of security testing?

How exactly shall AI be used to generate new test cases when testing AI? When is testing AI successful, what means "pass" for AI? Can we test AI by AI? How exactly can we use *Combinatory Logic* for testing AI?

How does ART fit into DevOps? Who shall prepare test cases and how shall test results be communicated to the user of software-intense systems?

CHAPTER 2: TEST METRICS

Today's software testing body of knowledge focuses on testing code. While testing code is important, testing the full system's functionality matters much more for the digital society. Code metrics, mostly captured by automated testing tools, are unfit for functionality tests for software-intense systems. Code is often unavailable for e.g., cloud services. Moreover, systems often use only parts of the total implemented functionality of some service. Then, testing the unused part does not matter.

Test metrics should refer to functionality, not code. This means that lines of code cannot be the reference for testing intensity; it must be functional size. For functional size, models exist that allow determining size at defined granularity for any service in use. The models also work for services that are only partially used. In the past, functional size models were used to predict cost of software projects and thus were not in the focus of the testing community. Testers were referring to code. Now, when testers face the challenge of digitalization, they must learn metrics for testing that are independent from code.

2-1 INTRODUCTION

Today's practices in software and system testing are quite strange. People count entries in bug inventories and mistake this count for the number of defects. Test cases refer sometimes to code and sometimes to the behavior of some unidentified piece of software. It remains unclear to what piece of software a bug report refers to.

Common testing techniques, metrics and tools refer to code – notwithstanding that code is often not available when testing software, and systems often rely on cloud services. Moreover, code is subject to the programming language, programming environment, and sometimes not even open. It is not possible to define any reasonable software metrics based on general code characteristics; you need always to be specific about what kind of code you want to measure for testing. How to test cloud services?

When consulting the ISO/IEC/IEEE 29119 testing standard (ISO/IEC/IEEE 29119-4, 2015), it astonishes that part 4 identifies 23 different so-called *Test Coverage Items*. This is already a step away from testing code. But tests primarily address software functionality. It is unnecessary to define extra "Items" to undergo testing.

Functional models are available and are used since the past 40 years for sizing software. Why shall test coverage items be something else than its model elements? Functionality of software can easily be assessed and modeled. While the availability of code is helpful because functional models can be generated automatically (Soubra, et al., 2014), in general *Functional User Requirements* (FUR) are enough. The only thing you need to know for testing software-based systems is what they are supposed to do. Since functional requirements exist not only for code written on purpose – e.g., user stories – for cloud services, or any standard software with proprietary code, they exist as well.

While *Non-Functional Requirements* (NFR) also exist and are testable as well, such tests can rarely be automated and are not considered in ART.

2-2 MODELING SOFTWARE

Any software can be modeled by its functional requirements. The ISO/IEC 14143 (ISO/IEC 14143-1:2007, 2007) defines what FUR exactly are and how to model them. They key statement is that model elements must cover everything that is needed to implement some FUR; thus, the ISO/IEC 14143 standard defines granularity. The level of granularity is defined by the user view represented in the FUR. Sometimes, general service considerations at the level of microservices are good enough; sometimes, code-level granularity is required, depending upon the *User* requesting a FUR.

The "U" in FUR is important: whatever functionality is modeled; it is important to identify its user. A user can be a human, another application service, or another layer in the system's architecture. Some lines of code might implement multiple FUR for different users; it is obvious that such a line of code can implement one or more FUR imperfectly, or completely faulty, while other FUR behave correct. This consideration alone demonstrates how misleading it is to link defects to code. To call today's practices in software and system testing strange, is probably not appropriate. With today's testing practice, it is a miracle that not more software fails than does today. The miracle is because software developers are perhaps the most responsible workers found today. Testers in turn often enough fail to help them effectively.

The lack of proper testing is a threat to ICT as a profession as well as all the economic churn put expectantly on ICT, digitalization for example.

2-2.1 METRICS FOR SOFTWARE

Before presenting the ways how to model software, let us introduce a related topic that somehow seems half-forgotten in the metrics community. Metrics is nothing new; since the beginning of civilization metrics have been indispensable for distributing goods, resources, wealth, and organizing welfare and warfare. Most people know the story how Eratosthenes calculated the size of the earth.

Citing Wikipedia: "Eratosthenes calculated the Earth's circumference without leaving Egypt. He knew that at local noon on the summer solstice in Syene (modern Aswan, Egypt), the sun was directly overhead. Syene is at latitude 24°05' North, near to the Tropic of Cancer, which was 23°42' North in 100 BC. He knew this because the shadow of someone looking down a deep well at that time in Syene blocked the reflection of the Sun on the water. He then measured the Sun's angle of elevation at noon in Alexandria … From these measurements, he calculated the angle of the sun's rays. This turned out to be about 7°, or 1/50th, the circumference of a circle. Taking the Earth as spherical, and knowing both the distance and direction of Syene, he concluded that the Earth's circumference was fifty times that distance."

Eratosthenes built a model that was not perfectly accurate but good enough for the purpose. He used a few simplifications, modeling the Earth as a perfect sphere, the sun rays as parallel, putting Alexandria due north of Syene. Then he could perform all necessary calculations on his model.

But how did Eratosthenes know the true distance between Alexandria and Syene? Pharaonic bookkeepers measured the distance between Syene and Alexandria regularly; an achievement that no civilization on Earth was able to repeat before France in the 18th century (Russo, 2004).

However, you cannot measure such a distance by foot or – at the time – by camel only in one go; you need to be able to measure parts of the distance and combine measurements correctly, using trigonometrical adjustments because the straight line is blocked sometimes.

This knowledge about the nature of metrics is the essence of the VIM and the GUM:

- The VIM: ISO/IEC Guide 99:2007, 2007. International Vocabulary of Metrology (ISO/IEC Guide 99:2007, 2007) – Basic and general concepts and associated terms (VIM);
- The GUM: ISO/IEC CD Guide 98-3, 2015. Evaluation of measurement data (ISO/IEC CD Guide 98-3, 2015) - Part 3: Guide to Uncertainty in Measurement (GUM).

Metrics must comply with the VIM and the GUM. Counting does not necessarily establish metrics. Counting points does not measure anything, unless the points mark units on a measurement scale.

2-2.2 MODELS FOR FUNCTIONAL SIZING

We observed that testing should not be against code alone but against functionality. For testing complex systems such as those powering autonomous vehicles, code is only partially available, and safety-impacting functionality depends as much from functions hosted in the cloud than from the local controls powered by embedded software.

For measuring tests, it is therefore straightforward to size tests based on models for the functionality of software. The size can be determined by counting model elements. Sizing tests against code is inappropriate. For describing functionality, FUR according ISO/IEC 14143 are the preferred kind of reference. Currently, four ISO standards exist that conform to the concepts of ISO/IEC 14143. From those, the ISO/IEC 20926 (ISO/IEC 20926:2009, 2009), for long years maintained by the International Function Point Users Group (IFPUG), is older and more widely used than all others.

2-2.3   THE IFPUG MODEL OF SOFTWARE

The IFPUG model (IFPUG Counting Practice Committee, 2010) defines a count for functional size by counting model elements that are conceptually familiar to traditional mainframe software: *Data Functions* and *Elementary Transactions*.

Figure 2-1: IFPUG Model – Three Transactions: EI, EO, EQ; Two Data Functions ILF, EIF

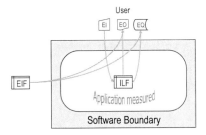

The IFPUG model identifies five elementary types of data functions or transaction, categorizing each model element as either low, medium, or high complexity, each with a fixed size number associated. These categories depend from the amount of data handled by each element, and the number of data references. Consequently, the categories define a jumping count.

Thus, with IFPUG, adding data elements can let the complexity assessment jump from one level into another. Or, in contrary, adding new elements to the model sometimes is not reflected in the count. Nevertheless, the IFPUG model can also be used to count *Test Points*, a test effort counting method for predicting test effort, proposed by Tom Cagley (Cagley, 2018). But there is no relation between test points and function points.

For knowing how to count model elements in ISO/IEC 20926, it is necessary to know the boundary for the complete system. The reason is that the total number of *Files Types Referenced* (FTR) impact the size of the transaction-type model elements. Without knowing the whole system, parts cannot be counted, according the IFPUG rules.

Consequently, the IFPUG count is not a metric; it does not conform to the VIM and the GUM. While it seems possible to adapt the IFPUG rules by allowing intermediate results instead of the jumps, and it is arguable that for practical purposes the FTR number is known well enough, namely from the transaction alone without regard to the whole system, such an enhancement of the IFPUG count towards a metric unfortunately is not yet on the agenda of the IFPUG counting committee. This makes the IFPUG counting method unattractive both for agile software development that needs to measure the functional size of sprints, and for test metrics.

2-2.4   TRANSACTION MAPS

The following Figure 2-2 explains with an example how to combine the model elements shown in Figure 2-1 to create a *Transaction Map*. Transaction maps are a way to visualize the IFPUG model for a software system. Depending upon the architecture, more than one transaction map might be needed. Then, typically one transaction map describes an application that manages an ILF, while others refer to the same elementary data element as an EIF. This in turn induces double counting for such elementary data functions that makes adding size for different applications unreliable

Figure 2-2: Transaction Map for the Navigator Piece of Software

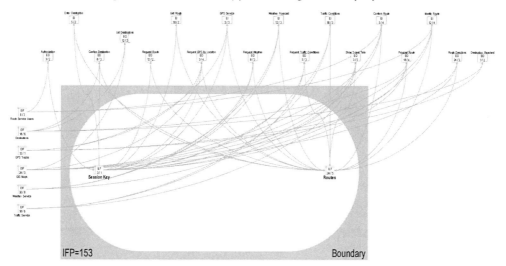

The *Navigator* application shown in Figure 2-2 is a piece of software using microservices such as GPS tracks, GIS maps, weather forecast and real-time traffic information to propose routes for a car; it is a simplified navigation system. The user can enter destination and the system proposes one or more possible routes to take, depending upon weather and traffic conditions. Favorite routes taken previously are not taken into consideration; routes are not attributable to car users, to protect their privacy. A session key is used to separate authentication from identification of the user.

Despite their failure to comply with the VIM and the GUM for sizing, transaction maps are ideally suited for use with agile teams, for visualizing which elements of software are touched in each sprint. The model elements are easily recognizable by businesspeople, somewhat less by developers; nevertheless, they can be used for communicating work done in sprints, at the same time providing its functional size.

Because of the said missing compliance to the VIM and the GUM, the sum of functional sizes delivered in sprints is significantly higher than the total functional size; however, this does not matter too much in practice. It rather reflects how the team struggles to get requirements right. Counting functional size with IFPUG is the better predictor for performance of an agile team that any other agile metrics, including story points. Anyhow, agile teams need be conscious about distinguishing new functionality from reused or enhanced existing functionality, to avoid unnecessary double counting in sprints.

The transaction maps like in Figure 2-2 serve many more purposes than just sizing functionality. Maps help to orientate and localize software elements. For tests, a map should be useful to localize test cases by identifying the model elements touched when running the test case. But with IFPUG, this is difficult. Test cases are not easily identifiable within a transaction map. The missing compliance to the VIM and the GUM create quite practical problems making it difficult to define test metrics based in the IFPUG method.

2-2.5 THE COSMIC MODEL OF SOFTWARE

In contrary to the IFPUG count, the COSMIC standard (ISO/IEC 19761:2019, 2019) complies to the VIM and the GUM. With COSMIC, you can measure parts, and from the parts one can construct the size of the compose. System boundaries are also defined in the standard but do not affect the count.

The COSMIC standard identifies layers. The layers' boundaries detect the flow of data moving from one object into another; however, the total count does not depend from how boundaries are drawn. Communication between functional processes require typically an Entry and an eXit, with a device in between that connects the two processes. Fortunately, devices and other applications yield the same count when crossing an application boundary.

A Read or a Write moves data between functional processes and persistent data stores. Every data movement transports a *Data Group*, identifying the data moved from one object to another. No application boundary is crossed.

A data movement moves a *Data Group*. Data groups hold the information needed to assess privacy protection needs, or safety risk exposure, of data. In certain cases, the data groups contain enough information to allow for generating code out of a COSMIC model (Oriou, et al., 2014).

Figure 2-3: The COSMIC Model, with Six Data Movements Entry, eXit, Write and Read

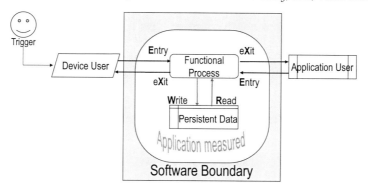

The constituent element of the COSMIC model is a *Functional Process*. A functional process is an object together with a set of data movements, representing an elementary part of the *Functional User Requirements* (FUR) for the software being measured, that is unique within these FUR and that can be defined independently of any other functional process in these FUR (COSMIC Measurement Practices Committee, 2017, p. 42).

Modeling software takes two distinct steps:

- Creating a model for the software is based on analyzing data movements and identifying the relevant objects of interest that are the origins and targets of such data movements. This step is called *Mapping* and results in uncovering the relevant functional processes.

- There must exist enough model elements to explain everything that is needed for the FUR. Create data movements only once per data group moved per functional process, notwithstanding how many times they are being executed.

Now you can count the *Functional Size* of software by counting the number of data movements. This is the way ISO/IEC 19761 COSMIC measures functional size. One data movement with a unique data group yields one *Function Point*. Two or more data movements moving the same data group between the same objects are considered only as one model element and, therefore, do not add to functional size.

2-2.6   DATA MOVEMENT MAPS

Data Movement Maps are a way to model a piece of software by connecting objects of interest, representing functionality, persistent stores, devices and other applications.

The connectors represent *Data Movements*. They have some resemblance to *UML Sequence Diagrams* (Bell, D., 2004) but with less detail; thus, without guards, loops, and alternative fragments. Also, sequencing is not prescribed.

Figure 2-4: Sample Data Movement Map

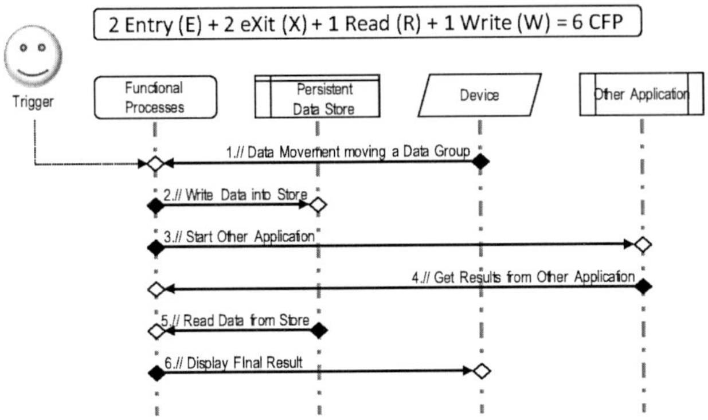

Data Movements always move a *Data Group*, which can be thought as a data record. Its uniqueness is indicated by color-filled trapezes. Another move of same data group between the same objects within a COSMIC functional process lets the trapeze blank.

2-2.7 OBJECTS OF INTEREST

For data movement maps, we distinguish four types of *Objects of Interest*:

- **Functional Processes**: Objects that perform functional processes in the COSMIC sense. One such object can perform several functional processes. Thus, such an object represents for instance one *Virtual Machine* (VM), or *Electronic Control Unit* (ECU) performing different calculations performing several functional processes in the sense of the COSMIC manual (COSMIC Measurement Practices Committee, 2017, p. 42).

- **Persistent Store**: Objects that persistently hold data. Contrary to the COSMIC definition, they provide data services to several different functional processes.

- **Devices**: a device can be a system user or anything providing data.

- **Other Applications**: other applications use functional processes the same way as devices do; however, they typically represent other software or systems that can be modeled the same way using data movement maps.

Triggers usually indicate the starting data movement of one COSMIC functional process. Thus, one functional process object can have several triggers.

2-2.8 THE NAVIGATOR APPLICATION AS COSMIC MODEL

Figure 2-5 models the *Navigator* application as a data movement map:

Figure 2-5: Data Movement Map for the Navigator Application

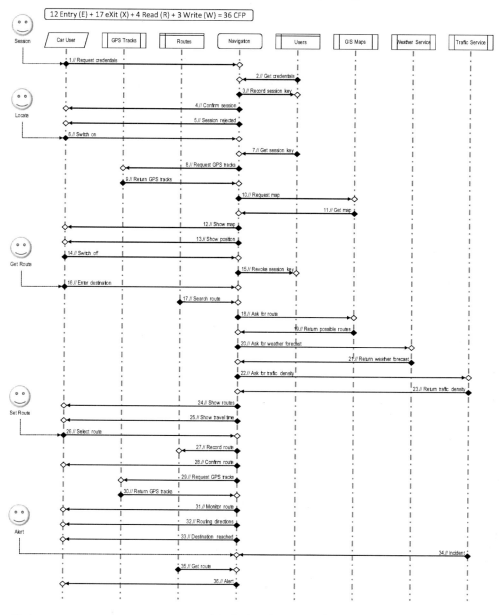

Data movement maps serve as graphical visualization of COSMIC models. The maps have lifelines just like UML sequence diagrams. The objects in a data movement map represent either functional processes, persistent store, devices, or other applications.

The difference to Figure 2-3 is that a lifeline belonging to a functional object can host more than one COSMIC functional process. Triggers are needed according COSMIC rules to initiate each functional process. As shown in Figure 2-5, more than one trigger can exist in a data movement map pointing to the lifeline of a specific object of interest representing functional processes. Thus, triggers pointing to an object of interest identify functional processes within this object.

Triggers also connect functional processes to user stories. A user story specifies a user triggering some functionality. This corresponds often to one COSMIC functional process; however, a user story might need more than one functional process to get completely implemented.

The *Navigator* application shown in Figure 2-5 consists of a mapping service, connected to routing and positioning, usually by the *Global Positioning System* GPS. Such an instrument is standard in today's cars, although we use a simplified model. It has the same functionality as the transaction map in Figure 2-2; however, the data movement map shows more details how the application technically works.

A route once chosen is used to tell the driver where to turn right, left, or around after missing the way completely. The *Navigator* relies on four external service applications: GPS, maps, weather, and traffic service. One functional object is enough; it uses two permanent stores, one for recording routes, the other one for recording users and their credentials. The relation between these two stores is critical for the ability of the Navigator to keep routes taken as private, in the ownership of the car user.

There are six different functional processes, all hosted by the *Navigator* functional object. The first functional process authenticates the user and creates a session, identifiable by a session key. The second uses that session key to locate the car on a map. The third functional process proposes routes to a destination chosen by the car user, based on weather and traffic conditions. The fourth functional process consists of selecting among different routes, if available, and storing the chosen route for further processing by the *Navigator* application, in case the *Traffic Service* application issues a traffic alert, or weather conditions change the expected travel time. This process continues with giving directions to the car user until the destination is reached. The fifth functional process informs the car user in case of a traffic incident that might cause choosing another route. The car user then must try another route.

There is no functionality provided here that for instance uses recorded routes to identify user preferences, avoiding privacy issues that arise from collecting routes chosen.

Counting the data movement yields 36 CFP since each data movement moves a different data group.

2-2.9 AUTOMATICALLY CREATING DATA MOVEMENT MAPS

The code given, creating automatically a model for functionality is easy for the COS-MIC model, for most programming languages. All one must do is identifying the objects of interest. In most programming languages, these are declared as objects. The only difficulty is to decide whether such an object is visible to the user and thus could correspond to a FUR. Then, once the objects are known, the data movements between the objects are easily identifiable. The effort is comparable to building a compiler.

Moreover, if different functional users can be identified, the same code may exhibit more than one data movement model. This is typical for a layered architecture, where the front-end functional user requires different data movements from different objects located in the middle layers or data layers, compared to what the end user requires from the front-end.

Sometimes, automatically creating a data movement map is without any extra effort. With today's *Microservice* architecture, constructing the model is directly possible from a *Kubernetes* network builder (The Kubernetes Authors, 2018). Kubernetes is a portable, extensible open-source container-orchestration platform for automating deployment, scaling and management of containerized applications. Kubernetes connects microservices by message pipes; for the respective granular view, this defines uniquely the functional size of a microservice architecture.

While IFPUG describes software functionality from a static viewpoint, COSMIC also addresses dynamic aspects. Since testing also is dynamic, COSMIC might be better suited for sizing tests than IFPUG also from that perspective.

2-2.10 STRENGTHS AND WEAKNESSES OF SOFTWARE METRICS

Counting Requirements. Software metrics count *Functional User Requirements* (FUR). This makes them independent from implementation details and allows comparing different solutions. Moreover, one does not need a finished product for counting its functional size. If functional size controls development cost, a functional count can be used to predict development cost. Also, it allows managing the scope of a project, e.g., for dealing with deadlines, and controlling budget.

Today's agile software development process has no clearly defined final set of requirements. Requirements are likely to change. However, as soon as there is a backlog, this is easy to count. You can track agile development by counting the backlog in sprints.

Repeatability. Software metrics should be independent from the actual counter. Counters are properly educated and certified for the method. A user community maintains a counting practice manual, and provides examination for certified professionals, making it easy to decide whether a count is correct or not.

The problem with that approach is that automatic counting is in principle not possible. Nevertheless, automatic counting methods exist but are approximations to the original manual count.

Independent from Implementation. Software metrics do not model implementation details. Functional size is the same for a single-user, closed application as for a mobile app using cloud services if the same is being calculated.

The concept of counting data movements in COSMIC matches the way modern software is build. Connecting *Docker* container service (Steve Singh et.al., 2018) can be modeled as a sequence of data movements between containers.

Independent from Algorithmic Complexity. Functional size models do not model mathematical algorithms, if they are not covered by the granularity of the FUR. Thus, if the FUR says, use some very complicated and computing-intensive algorithm that you can find maybe in a mathematical library, or you must implement manually, this does not add to functional size. Details and complexity of an algorithm does not impact functional size.

Independent from Non-Functional Requirements (NFR). Functionality does not depend neither from performance, nor from how much parallelism is implemented for load balancing. Software operated in single user mode counts the same as a service for many parallel users. Nevertheless, such NFR might in turn require additional functionality, by turning an NFR into FUR when looking at the respective granularity level. Performance improvements might require cache, and the functional cache user sees FUR and related functional size; load balancing also requires a load balancing functional process when looking at it from some internal layer. Functional size is indeed dependent from the viewpoint. This is the essence of the ISO/IEC 14143 international standard (ISO/IEC 14143-1:2007, 2007).

2-3 A SHORT PRIMER ON SIX SIGMA TRANSFER FUNCTIONS

Readers of the previous book by the author (Fehlmann, 2016) can skip this section, or quickly read through it as a refresher.

2-3.1 UNCOVERING HIDDEN CONTROLS

For decennials, *Quality Function Deployment* (QFD) is the discipline to uncover hidden customer needs for creating successful products (ISO 16355-1:2015, 2015). The main task is to capture the *Voice of the Customer* (VoC). Many proven methods and tools exist to understand the VoC and turn it into a prioritization profile.

QFD uses the concept of linear *Transfer Functions* in the form $y = Ax$, where y is the vector representing qualitative or quantitative user needs, and x the vector of quantitative parameters related to the technical solution. Since A is linear, it can be represented as a matrix (Fehlmann, 2003). It has many similarities to Six Sigma root cause analysis, where y is the observable response and A the matrix of measurements that correlate each vector dimension of x with each vector dimension of y. For measuring these correlations in Six Sigma, the *Design of Experiments* technique (Myers, et al., 2009) provides guidance how to get a sufficiently well-defined transfer function matrix for identifying main causes for an observed effect.

In both QFD and Six Sigma for manufacturing, finding the right controls for the vector x is the difficult part. Because of the non-decidability of first-order logic, there is no automated method possible to devise the "correct" instances of x, not even its dimensions – otherwise we would have a general problem solver and could let computers develop new technologies and new products. The main difference between Six Sigma in manufacturing and QFD is that, in QFD, proper measurements are often not possible. Classical QFD for product design replaces measurements by team consensus; thus, measuring expert judgment rather than physical evidence.

Measuring the response y in QFD involves techniques to understand the VoC that often rely on social science or involve not only mathematics but also psychology such as Saaty's *Analytic Hierarchy Process* (AHP) (Saaty & Alexander, 1989). Methods and techniques for the acquisition of the voice of the customer make up for the larger part of the ISO 16355 series of standards.

Finding the transfer function and assessing the right topics and dimension of x requires a very creative but disciplined process. This is the essence of QFD. As for any transfer function, it is possible to validate any pair of A and x by applying A to x. The result, Ax is a vector with the dimensions of the original response y, in QFD most often the voice of the customer. Because of the measurement errors and the uncertainty of expert judgements, $Ax \cong y$ but not equal.

The vector difference between Ax and y is called the *Convergence Gap*. This is an indication how well A and x together explain the response y, or in other words, whether a product or technology based on the quantitative parameters x and providing the transfer function A are capable to deliver the qualitative requested user needs y, thus validating the approach but not able to exclude the existence of other approaches.

Let x be the vector $x = \langle x_1, x_2, \ldots, x_n \rangle$, $y = \langle y_1, y_2, \ldots, y_m \rangle$ and $A = (a_{ij})$ the transfer function as a matrix, then the convergence gap is defined as the Euclidian distance between the m-dimensional vectors y and $Ax = \langle \sum a_{i1} x_i , \sum a_{i2} x_i , \ldots, \sum a_{im} x_i \rangle$:

$$\|\boldsymbol{y} - \boldsymbol{Ax}\| = \sqrt{\sum\left(y_j - \sum a_{ij}x_i\right)^2} \qquad (2\text{-}1)$$

The convergence gap can be used to optimize controls by iteration, using domain expertise, or by any other numerical optimization method. In fact, in Six Sigma the preferred method is the *Eigenvector Method* because it settles and flattens variations that originate from measurement errors or opinion blur, as observed by Saaty, and used for the *Analytic Hierarchy Process* (AHP) (Saaty, 2003).

2-3.2 THE HOUSE OF QUALITY

For decades, QFD has been identified with, and partially misunderstood as, the so-called *House of Quality* (HoQ). In the HoQ, the vector $\boldsymbol{y}$ is the profile of customer needs, as found by some suitable voice of the customer process, and $\boldsymbol{x}$ is the profile of the qualities required for the technical solution. Thus, QFD allows selecting optimum solutions, avoiding unnecessary gadgets that only add cost to the new product. For this, the HoQ is still ideal; however, the HoQ is only a small portion out of the QFD method. Nevertheless, it is the best-known part of the method, and popular among Six Sigma Black Belts and Marketing managers alike.

2-3.3 THE HELP DESK IMPROVEMENT EXAMPLE

For a HoQ example, assume, a Help Desk operator wants to improve its service. The help desk is a traditional one, with humans answering questions and helping customers who are not yet able to help themselves with the tools provided through the Internet. Humans sometimes can improve doing their jobs by receiving training, while machines undergo deep learning.

A simple pairwise comparison – a basic AHP session – identified the following priority profile $\boldsymbol{y}$ for a typical Help Desk customer:

Figure 2-6: Pairwise Comparison for the Help Desk House of Quality

AHP Priorities Customer's Needs	y1 Friendliness	y2 Responsiveness	y3 Accuracy	Weight	Ranking	Profile
y1 Friendliness	1	2	1	41%	1	0.69
y2 Responsiveness	1/2	1	2	33%	2	0.56
y3 Accuracy	1	1/2	1	26%	3	0.45

Profiles and weight follow the definitions used for AHP (Saaty & Alexander, 1989): the sum of the percentages is 100% while the profile represents a three-dimensional normalized vector of length 1, i.e., the sum of the squares of the coefficients yields 1, the unit vector length. From Saaty (Saaty, 1990) it is known that the AHP profile y of an AHP square matrix H is its *Principal Eigenvector*; thus, $Hy = y$ holds up to some limit of exactitude caused by the numerical algorithm. The eigenvector balances the inconsistencies out caused by human judgements in pairwise comparisons. Geometrically, an eigenvector points in a direction that is stretched by the transformation.

Profiles and weight percentages always transpose into each other. This is only a matter of convention. However, it is well known that you cannot add or subtract weight percentages, because this will no longer yield percentages, and even when recalibrating the result of addition, if the weights are out of balance, the resulting bias can become substantial. For comparing results from AHP, you must use profiles. Because of their nature as vectors, they allow addition and subtraction, and can be compared to each other, if they represent directions in a vector space only. The sum of two profiles yield another profile, as soon as normalized to length one.

Figure 2-7: The Priority Profile y for Customer Needs

Customer's Needs Topics	Attributes		AHP Priorities Weight	AHP Priorities Profile	
y1 Friendliness	Remains cool	Always friendly	41%	0.69	
y2 Responsiveness	Understands the problem	Finds a way to solve	33%	0.56	
y3 Accuracy	Complete information	Compelling	26%	0.45	

We investigate the following pair of quantitative parameters x and transfer function F for improving the Help Desk service:

A team of experts might now come up with the following *House of Quality* (HoQ):

Figure 2-8: The Transfer Function A (HoQ)

Critical To Quality Deployment Combinator		Goal Profile	x1 Training	x2 ICT Infrastructure	x3 Salary & Bonus	x4 Work Place	Achieved Profile	
Customer's Needs								
y1	Friendliness	0.69	9		2		0.67	
y2	Responsiveness	0.56		7		6	0.58	
y3	Accuracy	0.45	1		6	3	0.46	
Solution Profile for Critical To Quality:			0.65	0.41	0.41	0.49	Convergence Gap	
							0.03	

34 Total Effort Points
0.20 Convergence Range
0.20 Convergence Limit

The matrix correlates customer needs with effects originating from *Critical to Quality* controls. Solving the transfer function with the eigenvector method explained below (2-3.4) for the controls x yields an *Achieved Profile Ax* near enough to *Goal Profile y*.

For transforming the profile into percentages, consult Figure 2-9. Here the bottom profile of Figure 2-8 is turned by 90° to display horizontally.

*Figure 2-9: The Technical Solution Profile **x** – Critical to Quality*

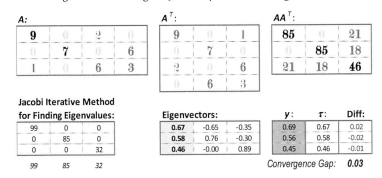

Critical To Quality Topics	Attributes			Weight	Profile	
				Priority		
x1 Training	Behavioral Training	With Stress Test	Must make fun	33%	0.65	
x2 ICT Infrastructure	Customer Identification	High Performance	High Reliability	21%	0.41	
x3 Salary & Bonus	NPS related	Predictable		21%	0.41	
x4 Work Place	Ergonomic	Individual	High Performance	25%	0.49	

This means that the following distribution will provide best value for money: by investing 33% of the total budget into *x1: Training*, 21% into *x2: ICT Infrastructure*, 21% into *x3: Salary & Bonus*, and 25% into *x4: Work Place*. The percentages indicate how the budget for improving the Help Desk services is allocated best.

Tradition restricted the cell values adopted in QFD transfer function matrices to 0, 1, 3, 9; with 9 as the highest correlation value. This was found suitable for expert team judgement; however, from a mathematical viewpoint any scale is permitted if the scale is a *Ratio Scale*; i.e., $9 = 3 \times 3$.

2-3.4 SOLVING A TRANSFER FUNCTION BY THE EIGENVECTOR METHOD

There are various mathematical or empirical methods available to solve $\mathbf{y} = A\mathbf{x}$, given the vector dimension of $\mathbf{x}$ and some matrix A. A cute way of solving is by using the AHP eigenvector method which has the advantage to flatten out measurement errors. Such errors are unavoidable especially if a team of experts is setting up the transfer function matrix. For this, we tilt the $m \times n$ matrix A over its diagonal into its $n \times m$ transpose A^T and multiply A with A^T; this yields a $m \times m$ positive-definite square matrix that has m Eigenvectors.

Figure 2-10: Solving the $\mathbf{y} = A\mathbf{x}$ problem with Eigenvectors

A:

9	0	2	0
0	7	0	6
1	0	6	3

A^T:

9	0	1
0	7	0
2	0	6
0	6	3

AA^T:

85	0	21
0	85	18
21	18	46

Jacobi Iterative Method
for Finding Eigenvalues:

99	0	0
0	85	0
0	0	32
99	85	32

Eigenvectors:

0.67	-0.65	-0.35
0.58	0.76	-0.30
0.46	-0.00	0.89

y :	τ :	Diff:
0.69	0.67	0.02
0.56	0.58	-0.02
0.45	0.46	-0.01

Convergence Gap: **0.03**

The solution relies on the theorem of *Perron-Frobenius*, saying that positive determined square matrices have a *Principal Eigenvector* τ which is all positive. For a short proof of this theorem, see e.g., Cairns (Cairns, 2014). The eigenvectors are calculated using the *Jacobi Iterative Method* (Volpi & Team, 2007), or any other suitable numerical solution method. Then, setting $x = A^\mathsf{T}\tau$ solves $Ax = A(A^\mathsf{T}\tau) = AA^\mathsf{T}\tau = \tau$, because τ is an eigenvector of the square, positive-definite matrix AA^T. If it happens that $\mathbf{y} \cong \tau$, i.e., the goal vector $\mathbf{y}$ is near enough to such an eigenvector τ of AA^T, the solution $\mathbf{y}$ is an approximate solution to the problem $\mathbf{y} = A\mathbf{x}$, up to the convergence gap.

2-4 MEASURING TESTS

A *Test* is a finite collection of test stories. *Test Stories*, in turn, are finite collections of test cases, characterized by the common business value they deliver. Test stories are often related to user stories but typically not the same. Test stories can address more than just one user story.

Test cases are represented as arrow terms, starting with a set of preconditions (test data) and yielding some response. In a data movement map, it is straightforward to identify those data movements that are executed if running a test case. The initial data movements are those whose data group last meets the assertions made on test data; the last data movement first meets the response assertion. Moreover, objects of interest can be expected to provide test stubs; this means that such objects can provide test data without executing all the data creation functionality that under normal operational conditions is needed. If there is some hardware in the loop, test stubs are needed anyway to simulate the sensors' or actuators' data supplied into the test.

2-4.1 TEST SIZE

Test Size thus is the minimal number of data movements needed to execute some test case to produce the test response. As with COSMIC in general, moving the same data group is counted only once for size. However, since a test story consists of many test cases, a specific data movement is executing many times within a test, typically with different test data. All test cases within a test story must be different from each other. Attributes contained within test cases must specify test data all different, otherwise the test cases are considered equal.

Test Intensity in turn is an average number characterizing how many times on average a data movement becomes part of test case. Since high test intensity does not rule out that not all data movements are executed at least once in a test, *Test Coverage* (see section 2-4.4) remains an important indicator, specifying the percentage of data movements not covered with one test case in some test story; see Figure 2-29: *Test Status Summary*.

The total size of a test story is the sum of all size of the test cases executed within a test story, thus increasing test size when executing more test cases.

In statistics, test distribution indicates the degree to which test intensity differs within one test story, or within the full test. For practical purposes, such a metric seems not very telling, since it does not replace test coverage. It is rather expected that high business value increases test intensity while data movements moving irrelevant data are well tested with a few test cases only. Thus, test intensity depends from business value and is not and is not normally distributed. Therefore, test distribution is not a meaningful indicator.

2-4.2 TEST WALK

The data movement maps can be used to visualize tests cases. You can walk the tests, similar, but less in detail, to walk through code. Such visualization might help in bulk testing for identifying bugs found. The tester sees selected sequences in the data movement map; he can "walk" the data movements when planning or executing tests. This makes functionality visible to the development team, localizes defects that impact functionality, and supports communication between testers, users, and developers. Figure 2-11 shows how *Data Walker* walks four data movements of a test case and detects a bug at the fourth data movement.

Figure 2-11: Test Walk on Data Movement Maps; one Bug Found in Forth Walk

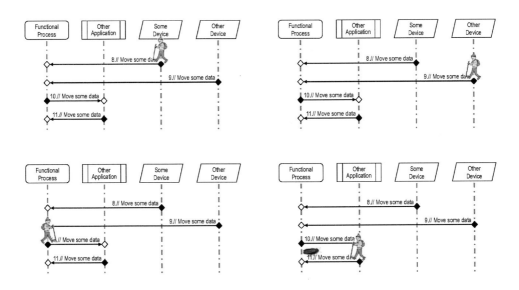

A *Bug* is defined the traditional way for testing: a test case that returns an unexpected response. Because our *Data Walker* can detect only one bug at a time, we are able to count defects unambiguously and thus define what defect density is. We count a maximum of one defect per data movement executed within a specific test story. The maximum number of defects per test case is its test size. Consequently, if the data walker detects bugs for different test stories in the same data movement, he can only count one defect per test story.

2-4.3   DEFECT DENSITY

A defect relates to requirements, specifications or expectation regarding the behavior of a system. If test cases are available, a defect means that the response does not meet the assertion of the response in the respective arrow term. It is therefore obvious that a defect relates to a test story. It refers to some data movement that exhibits the defect. Counting defects for each failed test case makes no sense if it refers to the same data movement.

Thus, counting defects become a limited task. You can count a maximum of one defect per data movement per test story. *Defect Density* is therefore a percentage of the total of defect opportunities. This definition opens the possibility to apply the usual Six Sigma techniques to defect density and defect distribution. Traditional defect counts obtained from counting the number of entries in a bug repository are not suitable for applying Six Sigma.

2-4.4   TEST COVERAGE

The key point for test metrics is *Test Coverage*. The problem with test coverage is that it has to do with users', or customers', values. It is useless to test pieces of software that deliver nothing visible to the user, or nothing that has any value. Test coverage has to do with FUR, with functionality, and nothing with code. Code implements functionality, and tests cover functionality, not code. Functionality can origin from anywhere, the cloud, other services. Code might provide other things that functionality.

For defining test coverage, functionality needs evaluation in view of customer values. It is obvious that just counting whether any given piece of functionality is covered by tests does not yield a useful metric, because users see value in respective functionality differently.

2-4.5   CREATING A CUSTOMER NEEDS PROFILE

The usual way of valuating functionality is by prioritizing user stories. Agile team set priorities when selecting user stories for a sprint; however, the methods used for setting priorities are not standardized. Since product owner is the most difficult role in agile development, especially with Scrum (Schwaber & Beedle, 2002), it is helpful to use a method dedicated to developing a product towards customer needs. The method is taken from *Quality Function Deployment* (QFD) (Fehlmann, 2016, p. 16).

The methods of choice for creating a customer needs profile is the *Analytical Hierarchy Process* (AHP), proposed by Saaty (Saaty, 2003) and used widely in various disciplines. AHP calculates eigenvector solutions; see Fehlmann (Fehlmann, 2016, p. 21). Our preferred alternative, combined with AHP, is the *Net Promoter® Score* (NPS) approach (Fehlmann & Kranich, 2014-2). The applicable ISO standard (ISO 16355-1:2015, 2015) lists many more excellent alternatives, e.g., (Mazur, 2014) and (Mazur & Bylund, 2009). Given that quality managers like surveys, net promoter is a method of evaluating surveys, avoiding large questionnaires by focusing on the *Ultimate Question* only (Reichheld, 2007). The ultimate question is how likely it is you would recommend a product or service to closely related persons. The second, related, question is why. Respondents are expected to provide a verbal statement. Then, it is possible to evaluate the responses by classifying answers into candidate business drivers – or customer needs – and calculating importance and satisfaction using transfer functions. The transfer functions try to uncover the customers' values for importance of, and satisfaction with, the candidate business drivers by trying to explain the observed NPS score with the solutions of the respective transfer functions. That will not always work but if it does, it is much more reliable than directly asking customers.

2-4.6   Effectiveness of the Implemented System

With customer needs established, user stories can easily be prioritized with a transfer function that maps user stories onto customers' needs. The transfer function uses the frequency of data movements needed for implementing the user stories. The resulting profile for the user stories can be used in agile development for prioritization.

In turn, mapping test stories onto user stories, again using the frequency of data movements used in test cases, defines *Test Coverage*, see section 2-4.4. The matrix looks familiar; tester use it to assess coverage of code by tests. But usually they are not aware of the convergence gap. If the test cases in a series of test stories cover the user stories, and the transfer functions yields a satisfactory convergence gap, this shows how well the test stories cover customer needs.

The test coverage matrix represents a transfer function providing assurance that the test stories verify the correct implementation of the user stories. The convergence gap is the metric that tells how well correctness can be proved by these tests.

Obviously, these tests do not prove anything else than the requirements expressed in the user stories have been correctly implemented. Adding user stories requires adding test stories. And as ever with transfer functions, there is no way of proving that the selected test stories are the only selection that works, not even the minimal one. The selected test stories work sufficiently well if the convergence gap closes. But that is enough for test automation, eliminating test stories that are not needed.

2-5 TEST METRICS FOR THE NAVIGATOR APPLICATION

Before continuing with theoretical statements, we look at a practical example: the navigation device application already encountered in *Section 2-2.8: The Navigator Application as COSMIC Model.*

2-5.1 CUSTOMER NEEDS, THE CAR USERS' VALUES

Customer needs are in our case rather the values of the car user, because it is unclear whether the car user is the same as the car owner and, even if so, if this is the direct customer of whoever offers the navigation device service. Also, car users are not necessarily car drivers; the car could drive autonomously.

We use two approaches:

- The Analytical Hierarchy Process (AHP)
- A Net Promotor Survey (NPS)

and combine the resulting profiles for the car users' values.

2-5.2 THE ANALYTIC HIERARCHY PROCESS

The AHP consist of pairwise comparisons between the following five potential values:

Figure 2-12: Car Users' Values

Customer's Needs Topics	Attributes			Weight	Profile	
				AHP Priorities		
y1 Find a Route	Fast	Secure	No jams	17%	0.37	
y2 Know Arrival Time	Reliable	Flexible		23%	0.51	
y3 Avoid Jams	Minimum traffic	Fast	Predictability	14%	0.31	
y4 Avoid Blockers	Incidents	Events	Bad weather	17%	0.36	
y5 Drive Safe	Road conditions	Avoid road works	Avoid populated areas	28%	0.61	

The navigation device cannot slow down a car if needed; that would be part of autonomous driving, or of an *Advanced Driving Assistance System* (ADAS) connected to the *Navigator.*

The AHP in Figure 2-13 puts the value *y5: Drive Safe* highest by assigning equal value as for *y1: Find a Route* but double the pairwise comparison weights against the other three proposed weights in the AHP matrix. The second in ranking is *y2: Know Arrival Time* which is obviously closely linked to value *y5.* However, this is difficult to find out by asking the user directly. The car user will rather pretend *y1*, *y3* and *y4*, finding the fastest route and avoiding jams and other blocking obstacles have highest priority.

Only pairwise comparison detects the true needs.

Figure 2-13: Analytic Hierarchy Process for Five Potential Car Users' Values

AHP Priorities Customer's Needs	y1 Find a Route	y2 Know Arrival Time	y3 Avoid Jams	y4 Avoid Blockers	y5 Drive Safe	0.23	Weight	Ranking	Profile
y1 Find a Route	1	1/3	1/2	2	1	0.07	17%	3	0.37
y2 Know Arrival Time	3	1	2	1	1/2	0.21	23%	2	0.51
y3 Avoid Jams	2	1/2	1	1/2	1/2	0.10	14%	5	0.31
y4 Avoid Blockers	1/2	1	2	1	1/2	0.21	17%	4	0.36
y5 Drive Safe	1	2	2	2	1	0.41	28%	1	0.61

2-5.3   NET PROMOTER® SCORE

Reichheld, Bain & Company, and Satmetrix Systems, Inc. have introduced and trademarked *Net Promoter® Score* (NPS) as a measurement method for customer loyalty (Reichheld, 2007). Because such considerations look somewhat odd, it is appropriate to ask the users of a car by means of a survey. Avoiding the useless direct question, we rather rely on the NPS methodology asking the car user whether he or she recommends our *Navigator* application, yielding the NPS score, and why she or he probably give this score – named the *Verbatim*.

The result looks as follows:

Figure 2-14: Response to NPS Survey by Three Segments of Car Users

Survey Results			Overall NPS:	25%	
Customer Segments	Attributes		NPS Profile	NPS	
NPS1 Business People	Meeting	Time Pressure Planned	0.61	29%	
NPS2 Professionals	Appointments Predictable		0.72	33%	
NPS3 Leisure	Shopping	Sightseeing Likes driving	0.33	15%	

A total NPS of 25% is nice but does not necessarily guarantee product success. For more detail on NPS, see (Reichheld, 2007), and for the methodology how to interpret it for VoC, see (Fehlmann & Kranich, 2012) and (Fehlmann, 2016, p. 104).

The verbatim responses were categorized into references to the five values listed in Figure 2-12. Counting the frequency of mention yields the importance given to these values; also considering the positive or negative value of the mention yields the satisfaction. Satisfaction can be used as a corrective to importance; however, since satisfaction be negative, namely dissatisfaction, it not always gives clear guidance on the relative importance of the five values. The method combines classification with counting.

Classification means to cluster words into notions described with these words and counting means simply to count how many times they appear in verbatims.

It should be noted that the term AI does not imply any of the concepts related to mindfulness, reasoning and understanding that other languages – such as German – connect to the terms derived from the Latin "*intellegere*". The Latin origin *intellegere* means read, or infer, between the lines, or other objects. Intelligence, in English, has a slightly other meaning. It is used to describe the activity of collecting data and turn it into knowledge by counting similarities found in such data. Secret *Intelligence Service* is exactly that. Artificial intelligence does not aim for reason, not even inducing appropriate behavior. But transfer functions can reveal the possible causes, even the most likely causes if used with due domain expertise.

We got the following two transfer function matrices:

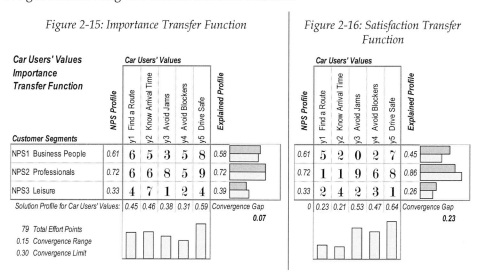

Figure 2-15: Importance Transfer Function   Figure 2-16: Satisfaction Transfer Function

Figure 2-15: Importance Transfer Function

Car Users' Values Importance Transfer Function / Customer Segments	NPS Profile	y1 Find a Route	y2 Know Arrival Time	y3 Avoid Jams	y4 Avoid Blockers	y5 Drive Safe	Explained Profile
NPS1 Business People	0.61	6	5	3	5	8	0.58
NPS2 Professionals	0.72	6	6	8	5	9	0.72
NPS3 Leisure	0.33	4	7	1	2	4	0.39
Solution Profile for Car Users' Values:		0.45	0.46	0.38	0.31	0.59	Convergence Gap 0.07

79 Total Effort Points
0.15 Convergence Range
0.30 Convergence Limit

Figure 2-16: Satisfaction Transfer Function

	NPS Profile	y1 Find a Route	y2 Know Arrival Time	y3 Avoid Jams	y4 Avoid Blockers	y5 Drive Safe	Explained Profile
NPS1 Business People	0.61	5	2	0	2	7	0.45
NPS2 Professionals	0.72	1	1	9	6	8	0.86
NPS3 Leisure	0.33	2	4	2	3	1	0.26
	0	0.23	0.21	0.53	0.47	0.64	Convergence Gap 0.23

The value *y5: Drive Safe* wins again; however, the second rank is not so clear. Obviously, satisfaction is high with the ability of our *Navigator* to avoid jams.

Combining importance and satisfaction transfer function profiles for the car users' values yields:

Figure 2-17: Combining Importance and Satisfaction from the NPS Survey

Combined from NPS Survey

Car Users' Values	Attributes			Importance 5	Satisfaction Gap 1	Σ	NPS Priority Weight	NPS Priority Profile	
y1 Find a Route	Fast	Secure	No jams	2.23	0.60	2.63	22%	0.48	
y2 Know Arrival Time	Reliable	Flexible		2.30	0.63	2.93	22%	0.49	
y3 Avoid Jams	Minimum traffic	Fast	Predictability	1.89	0.29	2.18	17%	0.37	
y4 Avoid Blockers	Incidents	Events	Bad weather	1.56	0.34	1.89	15%	0.32	
y5 Drive Safe	Road conditions	Avoid road works	Avoid populated areas	2.95	0.22	3.18	24%	0.54	

The *Satisfaction Gap* is useful as a corrective. The satisfaction gap weights negative statements exponentially; it thus stretches the importance profile in case of dissatisfaction. If customers are dissatisfied with an unimportant topic, the satisfaction gap remains nevertheless small and does not affect the profile (Fehlmann, 2016, p. 117).

In our case, the ranking is almost the same as with the AHP. Since satisfaction has not been very reliable, looking at the convergence gap, it is considered as a corrective only, with weights five (5) against one (1), in favor of the importance profile and ranking (Figure 2-17).

2-5.4 VOICE OF THE CUSTOMER

There exist many more methods to measure the *Voice of the Customer* (VoC). Among these are such simple things as voting. We can also draw a vote amongst car users' what matters to them most, and a possible result could be as shown in Figure 2-18:

Figure 2-18: Sample Vote of Car Users on their Values

Customer's Needs Topics	Attributes		VoC Input	Weight	Profile	
y1 Find a Route	Fast	Secure	16	19%	0.36	
y2 Know Arrival Time	Reliable	Flexible	0	0%	0.00	
y3 Avoid Jams	Minimum traffic	Fast	34	41%	0.76	
y4 Avoid Blockers	Incidents	Events	21	25%	0.47	
y5 Drive Safe	Road conditions	Avoid road works	12	14%	0.27	

Combining AHP, NPS, and VoC car users' profiles yields:

Figure 2-19: Combined Profile from AHP, NPS, and VoC for Car Users' Values

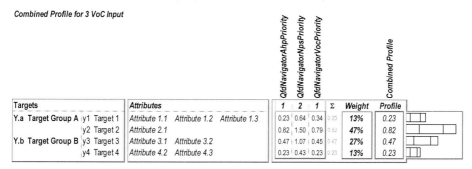

Here, in Figure 2-19, the NPS survey has been given double the weight than the AHP and the VoC, because NPS did ask more people at once than AHP or the VoC survey.

2-5.5   THE USER STORY PROFILE – FUNCTIONAL EFFECTIVENESS

With help of the car users' profile, the user stories can easily be prioritized by help of a transfer function. The transfer function for *Functional Effectiveness* originates from the data movement map. Do the data movements cover all needs of the customer, as expressed by the FUR, or user stories?

Functional effectiveness is easily measurable; it simply means assessing which data movements contribute to what goal target, and then compute the convergence gap. A software is functionally complete and effective, if the convergence gap closes.

Functional effectiveness has practical value. While missing functionality hints at missed business values, sometimes functionality is required that does not contribute to some of the values; maybe other reasons call for it. Then the convergence gap closes only if those other requirements are part of the value profile.

Figure 2-20: User Stories for the Navigator Application

	User Stories Topics	As a ... [functional user]	I want to ... [get something done]	such that ...[quality characteristic]	so that ... [value or benefit]
1)	Q001 Authentication	Car User	authenticate myself	I can use the Navigator	I remain anonymous for the Navigator
2)	Q002 Get Route	Car User	get the fastest route	I arrive at the predicted time	I can make arrangements for work and leisure
3)	Q003 Safe Route	Car User	arrive safely	the predicted driving time remains valid	I arrive at the predicted time
4)	Q004 Avoid Jams	Car User	use a route around traffic jams	I arrive at the predicted time	I can make arrangements for work and leisure
5)	Q005 Avoid Storms	Car User	avoid bad weather conditions	I arrive at the predicted time	I can make arrangements for work and leisure
6)	Q006 Use Routes	Car User	know my Driving Assistant where to go	I can use it without hesitation	the Driving Assistant knows where to go
7)	Q007 Locate	Car User	know my position	I know where I am	the Navigator can calculate travel time
8)	Q008 Set route	Car User	decide which route to take	I can exhibit my preferences	the car takes my preferred route
9)	Q009 Navigate	Car User	know which direction to go	I can rely on my Navigator	I reach the destination directly

Thus, the question is interesting in both cases: why some software is functionally effective or not. In practice, checking for functional effectiveness is a means to detect both missing functionality and excess functionality; consequently, it is a metric of high interest for Lean Six Sigma practitioners.

Figure 2-21: Get Route supporting y1: Find a Route

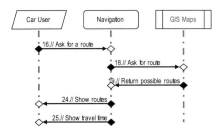

To assess functional effectiveness, it suffices to count how many data movements support some specific car users' value. However, such an assessment is not straightforward; sometimes it can be disputed whether a data movement carries specific importance for one of the car users' values. Since we use that information later for test coverage, the importance should be derived from the criticality of proper functioning of such data movement.

The technique used for identifying such data movements is extracting the user stories from the data movement map. E.g., from the *Navigator* map in Figure 2-5, the *Q002: Get Route* user story supports the *y1: Find a Route* value for the car user with the following five data movements, see Figure 2-21.

Figure 2-22: Functional Effectiveness for the Navigator Application

Car Users' Values	Goal Profile	Q001 Authentication	Q002 Get Route	Q003 Safe Route	Q004 Avoid Jams	Q005 Avoid Storms	Q006 Use Routes	Q007 Locate	Q008 Set route	Q009 Navigate	Achieved Profile
y1 Find a Route	0.41	8	5				5	9	4	2	0.40
y2 Know Arrival Time	0.56		6	5	6	6	2	2	2	8	0.53
y3 Avoid Jams	0.36		3	6	5		4	4		3	0.38
y4 Avoid Blockers	0.41		3	6	4	5	4	4			0.38
y5 Drive Safe	0.47	3		6	7	5		5	5	5	0.52
Solution Profile for User Stories:		0.19	0.31	0.42	0.42	0.32	0.25	0.42	0.21	0.36	Convergence Gap 0.06

157 Total Effort Points
0.10 Convergence Range
0.20 Convergence Limit

Doing that for all combinations of user stories and car users' values yields the *Functional Effectiveness* transfer function (Figure 2-22) with a convergence gap of 0.05;

- 42 -

indicating that the *Navigator* application is indeed a valuable, efficient and effective implementation of the car users' need for a valid navigation device. That means, it provides enough functionality, but no excess functionality without value for the customer.

The cells in the functional effectiveness transfer function count the number of data movements supporting each of the car users' values. Since the application has 36 CFP only but the total count – called *Effort Points* – is 157, it is obvious that many data movements support more than just one of the five car users' values. This is a sort of classification we need for later applying AI to automate testing.

2-5.6 TEST COVERAGE FOR THE NAVIGATOR APPLICATION

Creating test stories covering the user stories for the *Navigator* application is rather straightforward, based on the few user stories selected to fit into a book.

Figure 2-23: Thirteen Test Stories for the Navigator Application

Test Stories for Capability Testing

Test Cases

	Test Story	Case 1	Test Data	Expected Response	Case 2	Test Data	Expected Response
A Identity	A.1 Session Key	A.1.1	{User in good standing, User known}	Session key issued	A.1.2	{User didn't pay, User known}	Session key denied
	A.2 Session Ends	A.2.1	{Session key valid}	Session key revoked	A.2.2	{Session Timeout}	Session key revoked
	A.3 User Identity	A.3.1	{Match session key with user data}	No match	A.3.2	{Login user twice}	Session Key issued
B Routing	B.1 Destination	B.1.1	{Valid destination}	Route proposed	B.1.2	{Invalid destination}	Destination rejected
	B.3 Shortest	B.3.1	{No obstacles, route is free, weather fair}	Shortest route proposed	B.3.2	{Traffic jam detected}	Bypass proposed
	B.4 Safest	B.4.1	{No obstacles, route is free, weather fair}	Avoids populated areas	B.4.2	{Bypass proposed}	Avoids populated areas
	B.5 Obstacle	B.5.1	{Build-up of traffic jam}	Alert!	B.5.2	{Storm detected}	Alert!
	B.6 Alternate	B.6.1	{Alternate route recommended}	Alternative proposed	B.6.2	{No alternative available}	Inform
	B.7 Incident	B.7.1	{Sudden traffic obstacle}	Alert!	B.7.2	{Incidence ahead}	Ask destination
	B.8 Select	B.8.1	{Proposed routes, travel times, alerts}	Ordered proposals	B.8.2	{Select route}	Show chosen travel time
C Navigate	C.1 Direction	C.1.1	{Arriving at crossing}	Show direction	C.1.2	{Traffic jam detected}	Alert!
	C.2 Track	C.2.1	{On map}	Show position	C.2.2	{Lost GPS}	Alert!

Figure 2-24: Thirteen Test Stories for the Navigator Application (cont.)

Test Stories for Capability Testing

	Test Story	Case 3	Test Data	Expected Response	Case 4	Test Data	Expected Response
A Identity	A.1 Session Key	A.1.3	{User unknown}	User redirected	A.1.4	{Switch on}	Show map & position
	A.2 Session Ends	A.2.3	{Try credentials more than 3 times}	Session key denied	A.2.4	{Switch off}	Route deleted
	A.3 User Identity	A.3.3	{Exchange session key}	Session ends	A.3.4	{2nd session}	Both continue
B Routing	B.1 Destination	B.1.3	{Match List, Completed Entry}	Valid destination	B.1.4	{Session expired}	Get new session key
	B.3 Shortest	B.3.3	{Storm detected}	Bypass proposed			
	B.4 Safest	B.4.3	{User preference}	According preferences			
	B.5 Obstacle						
	B.6 Alternate						
	B.7 Incident	B.7.3	{Incident, Request driving track}	Activity track			
	B.8 Select	B.8.3	{Select route}	Show on map			
C Navigate	C.1 Direction	C.1.3	{Destination not set}	Show map only			
	C.2 Track	C.2.3	{User track}	No user found	C.2.4	{Switch off}	Revoke session key

For instance, the four test cases for test story *B.1: Destination* are:

- B.1.1: {Valid destination} →
- B.1.2: {Invalid destination} →
- B.1.3: {Match List, Completed Entry} →
- B.1.4: {Session expired} → Get new session key

The third case refers to an entry completed by matching destinations from a list. The corresponding data movement maps are:

Figure 2-25: Test Case B.1.1: {Valid destination} → Route proposed

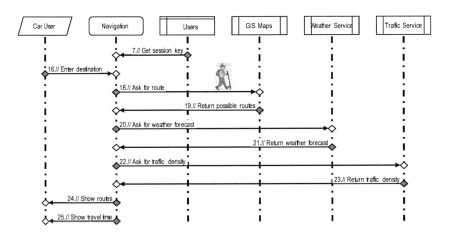

Figure 2-26: Test Case B.1.2 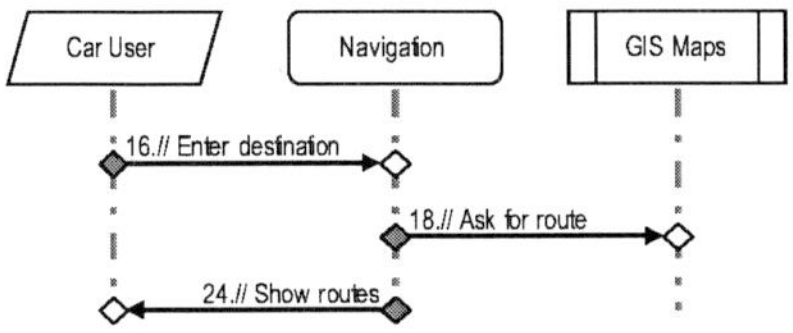   *Figure 2-27: Test Case B.1.3*

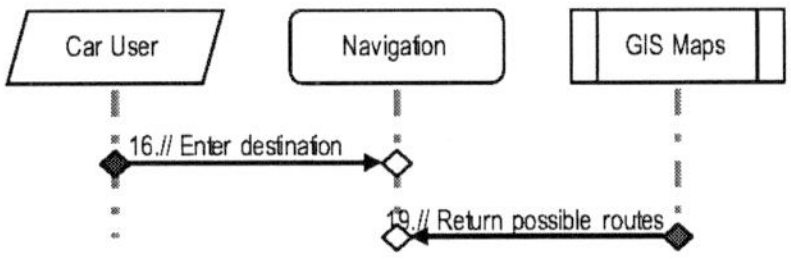

We visualize test flow by letting a *Data Walker* walk data movements, for instance in Figure 2-25, and count how many bugs he encounters; he's allowed to count only one bug per data movement and test story. Thus, he classifies data movements into those executing a test story correct, and those moving faulty data. This rule limits the total number of defects within an application that can be found by testing.

The two smaller test cases in Figure 2-26 and Figure 2-27 use only a part of the data movements needed to propose a route. The last one (*B.1.3*) tests a part of the process of entering a destination. Entering a destination shall be made easy by completing partial entries of a destination's name. For instance, you can enter the two or three initial characters of a valid destination and press the *Enter* key – or close entering data by any means suitable for the input device used – and the system will select the unique match from a list of valid destinations known to the *GIS Maps* application or, for more than one match, propose selecting from all valid matches. This ease-of-use functionality is considered part of "all that is needed to complete the '*16) Enter destination*' data movement". Our data walker on Figure 2-25 has just left this data movements, after checking with *7) Get session key*, continuing on *18) Ask for route* searching for defects.

Figure 2-28: Test Coverage Transfer Function Showing Good Test Coverage

User Stories	Goal Test Coverage	1) A.1 Session Key	2) A.2 Session Ends	3) A.3 User Identity	4) B.1 Destination	5) B.2 Shortest	6) B.3 Safest	7) B.4 Obstacle	8) B.5 Alternate	9) B.6 Incident	10) B.7 Select	11) C.1 Direction	12) C.2 Track	Achieved Coverage
Q001 Authentication	0.19	11	12	30	6		4			4			3	0.14
Q002 Get Route	0.31		1		15	7	9	6	9	1	11	4	1	0.22
Q003 Safe Route	0.42	1	3	4	13	13	10	16	12	11	19	17	7	0.45
Q004 Avoid Jams	0.42	3	4	6	8	14	11	11	11	15	14	18	8	0.42
Q005 Avoid Storms	0.32	3	4	6	8	12	11	10	7	11	12	13	8	0.35
Q006 Use Routes	0.25		1		8	8	9	7	9	8	11	9	3	0.26
Q007 Locate	0.42	8	6	16	8	10	15	4	8	8	17	12	11	0.38
Q008 Set route	0.21		1		5	8	9	7	9	8	10	9	3	0.25
Q009 Navigate	0.36				3	7	6	17	6	19	9	25	9	0.38
Ideal Profile for Test Stories:		0.08	0.10	0.19	0.26	0.31	0.31	0.31	0.27	0.34	0.40	0.44	0.21	Convergence Gap 0.12

854 Total Test Size
0.15 Convergence Range
0.25 Convergence Limit

Consequently, it is possible to count all data movements that belong to test story *B.2: Shortest*. The total count is 79 – the sum of its column in Figure 2-28; however, only 7 of these data movements aim at user story *Q002: Get Route*.

The resulting test coverage matrix in Figure 2-28 has a favorable convergence gap of 0.12.

Again, the cells of the matrix contain the frequency of executing data movements by the test stories. We use the knowledge from *Functional Effectiveness* for assigning data movements to user stories in the row of the matrix. Thus, the test coverage matrix results from the selected test cases automatically; no further assessment of data movements is needed.

Real-world test coverage matrices have the dimensions of the number of user stories: a few hundred up to thousands, and test stories typically even more than user stories. Automatically generated test coverage matrices, measured with the convergence gap, are indeed indispensable for making the approach feasible and attractive.

Real-world applications also have a few hundreds to several thousand CFP functional size; thus, without machine-collectable data, and automated testing, test metrics remain theoretical stuff.

The test statistics for our *Navigator* application looks as follows:

Figure 2-29: Test Status Summary for Navigator

Total CFP:	36	Test Size in CFP:	854
		Test Intensity:	23.7
Defects Found in Total:	0	Defect Density:	0.0%
Defects Pending for Removal:	0	Data Movements Covered:	100%

2-5.7 KEY FIGURES FOR TESTING

The total *Test Size* depends from the number of test stories in place, as typically every data movement is tested several times in view of other FUR, or user stories. It counts how many data movements are executed by test cases, in total.

The *Test Intensity* (see section 2-4.1) tells how many times in average. This is test size divided by functional size; its dimension is the ration between functional size, in CFP, and test size, also in CFP. Thus, it is dimensionless.

The percentage of data movements covered by tests is what used to be called *Test Coverage*; however, test coverage is a matrix, not a key figure. The key figure that matters indicates *Data Movements Covered*; it is in memory of the traditional *Code Lines Covered* by tests that is still in use with testers, although it is not a metric and meaningless especially for cloud services.

In any case, *Defect Density* should be zero for safety-critical software, or near to zero in all other cases. Real-world applications are likely not to remain without defects; nevertheless, users would dearly like to know how many. Statistical methods exist to predict the residual defect density after the testing process; nevertheless, predictions are not actual measurements. The important point with defect density measured by COSMIC according ISO/IEC 19651 is, that the total number of possible defects is known, considering that defects count only once per data movement and per test story.

Consequently, it is well known when a software is so buggy that every data movement is faulty; also, if it passed all tests without a single bug detected. However, even in this favorable case, adding more test stories might result in detecting previously undetected bugs. Because of the test coverage transfer function, this is likely to cause more user stories to appear; that is, new functionality added to a software causes new defects to appear, even in well-tested code. This is the reality developers experience; users and customers rather find it difficult to understand why their need for such functionality has not been detected much earlier.

However, key figures are not here to express feelings, or frustration. They shall reflect the reality, and for this reason we need to say goodbye to the familiar pseudo-metrics used in software testing, still declared as best practices nowadays, see (ISTQB, 2011) and (ISTQB, 2014).

2-5.8   DEMING CHAIN OVERVIEW FOR TESTING

The most important precondition for automated testing is to know the goals of testing. Without the goals there is no way to help a robot or algorithm to decide whether it does the right kind of testing.

Figure 2-30: Deming for Tests

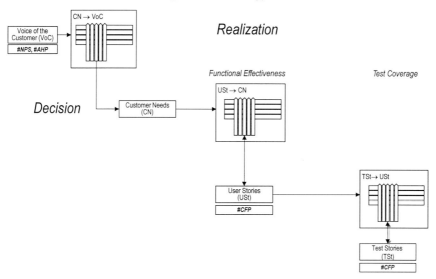

The *Deming Chain* shown in Figure 2-30 might serve as graphical overview for the method used.

2-5.9 Autonomous Real-time Testing for the Navigator?

There is not much interaction of the navigator with the real world. GPS delivers the location on a map, but the map is not maintained by the Navigator application. The map changes over time; also, road construction sites impose new obstacles, but all this is not done within the navigation device. Therefore, there is little to test after release, and nothing that cannot be tested when releasing updates.

Safety by a navigation device is not a big concern. Respecting *Privacy* is somewhat more challenging: while tracking cars is important for the *Traffic Services* application for predicting jams and detecting obstacles, such tracks should remain anonymous. Identifying car users might be useful for personalized advertisement based on the geographical location; however, for this navigation services is less useful than other devices such as a smartphone that can point its user immediately to shops and attractions. It is therefore safe to assume that privacy violations by navigation devices is rather limited and not subject to change over time.

Nevertheless, privacy checks during the operating lifetime of a navigation system may at least prove the validity of such an assumption.

2-6 Conclusion

Test metrics are of low interest for consumers that do not care for any risk connected to software. The *Navigator* is an example of an interconnected software-intense system that has no immediate need for more testing after released to the public. Even real-world larger-size systems with more than just skeleton functionality do not pose threats to safety, and rather few for privacy. Sharing routes taken, after all, is what most people gladly do without hesitation.

Nevertheless, under certain special conditions people do not like to share location and routing to everyone. In this case, privacy protection might become essential even for a simple navigation service.

The need for consumers to understand how well their privacy is protected exists even for such harmless services, and if consumers do not care, then it is because they fail to understand the impact of big data and the ability of AI-driven software to steal their privacy.

The fourth chapter exhibits a general proposal how privacy protection, and safety risk exposure, shall be made visible to the public. However, before that we look at the *Internet of Things* (IoT) requiring ART.

CHAPTER 3: TESTING THE INTERNET OF THINGS

The Internet of Things (IoT) has become very famous recently and a break-through is expected when the new 5G standards in mobile internet coverage become widespread. Testing the IoT meets the challenge that the system under test is unstable; simply, because it is extensible. You can always add another intelligent thing to an existing IoT concert, and thus expand the system.

How do you test expandable software systems?

3-1 INTRODUCTION

Combinatory Algebra, investigated by Erwin Engeler since the Seventies of the last century (Engeler, 1995), is the mathematical theory of choice for automatically extending test cases from a simpler, restricted system, to test stories that fully cover a larger, expanded system. The extension works only if software testing not only is automated but measured. Metrics must be independent from current implementation and from actual system boundaries.

Metrics for testing are based on the international standard ISO/IEC 19761 COSMIC.

3-1.1 METHODOLOGY

Figure 3-1 shows a data movement map for a simple data retrieval application, with a total functional size of 5 CFP according ISO/IEC 19761 COSMIC (ISO/IEC 19761:2019, 2019).

Figure 3-1. A Data Movement Map for Data Retrieval

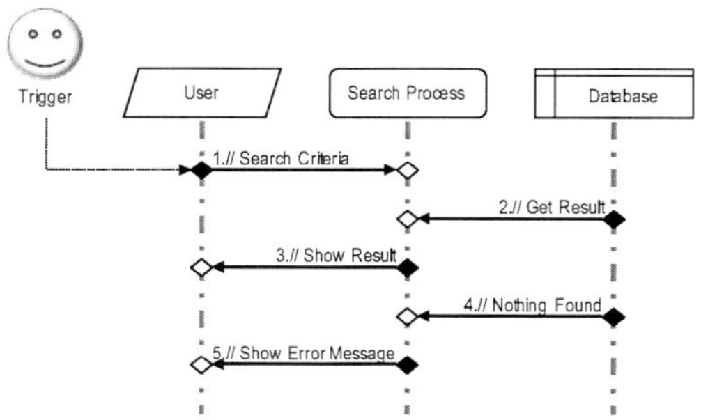

The map identifies three objects of interest – a user device, a functional process for search, and a persistent data object – and the data movements (or UML messages) between them. The data movements' count represents the functional size of an application. The number of data movements moving a unique data group determines functional size in *COSMIC Function Points* (CFP). The exact conditions when and how to count data movements according ISO/IEC 19761 is documented in the COSMIC measurement manual (COSMIC Measurement Practices Committee, 2017).

3-1.2   REAL-TIME TESTING

Real-time testing is the process of testing real-time systems and its software (Ebner (Ebner, 2004)). Real-time does not mean "testing anytime", but it means tests in limited time, within a freely selectable and adjustable time frame. Usually, real-time testing occurs while the battery loads, or the device is parked and restful.

The theory of *Combinatory Logic* postulates the existence of *Combinatory Algebras* whose computational power is Turing-complete (Barwise, et al., 1977); i.e., all programs that are executable by computers can be modeled. This guarantees the best achievable test coverage.

With combinatory algebra, test cases extend from real-time tests, covering a base system, to the actual, expanded system.

3-1.3   AUTONOMOUS TESTING

Autonomous testing is automated testing; however, without the need of simultaneous presence of a responsible test manager, or tester. The system executes tests autonomously, by connecting to some test case database, downloading the test cases as needed, executing the tests, and recording responses.

This requires the software be equipped with test stubs capable of accessing the test case database, and able to supply test data instead of a user device, or another application that accesses the system under test.

Test stubs can be present in any object; however, most test stubs reside in device and application objects. Such a system of test stubs replacing actual sensors, actuators and other hardware-in-the-loop are called *Digital Twins*. A digital twin refers to a digital replica of potential and actual physical assets (physical twin), processes, people, places, systems and devices that can be used for various purposes. For a recent discussion of digital twin's technology, see El Saddik (El Saddik, 2018).

3-2 TESTING THE INTERNET OF THINGS (IOT)

The *Internet of Things* is a collection of sensors, actuators, and services that connect hardware elements to software that reacts on events or collects data for further analysis. Such services are often hosted in some cloud, and the term *Web of Things* commonly refers to this. The IoT impacts the physical world over actuators, such as motors, locks, braking and steering controls.

The IoT changes scope and behavior with every sensor added or removed. Autonomous cars are a relatively simple example of an IoT within a container; as soon as they start talking to each other, for instance to find out where the other approaching car is heading to, the scope of the IoT is changing. Smart homes are intrinsically more complex since they are subject to external controls such as power plants optimizing the power supply over time.

Most IoT components remain small and tiny and have no great complexity by themselves. A temperature sensor reports actual temperatures on a continuous but limited scale; an actuator might lock doors or continuously dim light as needed. Their state is relatively easy to describe by terms over the physical world, called *Assertions*. Assertions describe test cases and test responses. This is an immediate application of combinatory logic.

Test cases have the structure of arrow terms (1-1). The arrow terms represent tests; in $a_i \rightarrow b$, the a_i describe the test data and b the test response. Responses can be as simple as the amount of impact on the actuators in an IoT orchestra.

The necessity for test cases produced automatically in IoT is apparent. There are no testers present when users connect a new sensor to their smart home network, or two autonomous cars with different software versions meet each other for the first time. Behavior of the newly connected system still must remain safe.

3-2.1 A SIMPLE IOT TESTING CASE

The mechanism in place are shown with a simplified IoT network. Consider a simple data retrieval application. The application meets two functional (FUR) and two non-functional (NFR) requirements with the following goal profile. The requirements and their profile represent *Customer Needs*, found by suitable *Voice of the Customer* techniques, see Figure 3-3. For our sample IoT application, we call them *IoT Needs*. the method of choice for profiling them is the *Analytic Hierarchy Process* (AHP; Figure 3-2).

Figure 3-2: Analytic Hierarchy Process for IoT Needs

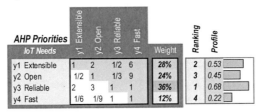

AHP Priorities IoT Needs	y1 Extensible	y2 Open	y3 Reliable	y4 Fast	Weight	Ranking	Profile
y1 Extensible	1	2	1/2	6	28%	2	0.53
y2 Open	1/2	1	1/3	9	24%	3	0.45
y3 Reliable	2	3	1	1	36%	1	0.68
y4 Fast	1/6	1/9	1	1	12%	4	0.22

For an explanation of the AHP and the tool used here to calculate, see (Fehlmann, 2016) or the original literature, e.g. (Saaty, 2003).

Figure 3-3: IoT Needs Priority Profile

	IoT Needs Topics	Attributes			AHP Priorities Weight	Profile
FUR	y1 Extensible	Easy to extend IoT	Device independent	Flexible	28%	0.53
	y2 Open	Open Source	Open Interfaces		24%	0.45
NFR	y3 Reliable	Always correct	Always secure	Safe	36%	0.68
	y4 Fast	No waiting			12%	0.22

Only three user stories are needed to cover these requirements:

Figure 3-4: User Stories covering IoT Needs

User Stories Topics	As a … [functional user]	I want to … [get something done]	such that …[quality characteristic]	so that … [value or benefit]
1) Q001 Search Data	Search Data App User	find data matching my search criteria	It's attractive	I know when data exists
2) Q002 Answer Questions	Search Data App User	know whether some data exists	answers are correct	I know when data doesn't exist
3) Q003 Keep Data Safe	Search Data App User	make sure my data is safe	it cannot be deleted	I can retrieve it if necessary

For user stories, we use the four-tailored *Fagg & Rule* form, see (Fehlmann, 2016, p. 158). The data movement map in Figure 3-1 with five data movements implements these three user stories.

This yields the following priorities for user stories, see Figure 3-5:

Figure 3-5: User Stories' Priority Profile for Simple Data Retrieval

User Stories Topics	Priority Weight	Profile
1) Q001 Search Data	32%	0.55
2) Q002 Answer Questions	40%	0.68
3) Q003 Keep Data Safe	29%	0.49

The user stories priority profile is a consequence of the customer needs profile in Figure 3-3. The total functional size according ISO/IEC 19761 COSMIC is 5 CFP, i.e., six data movements only; thus, this is a very small and simple application. The user stories' profile reflects IoT Needs as shown in Figure 3-3 by transfer functions (see section 2-3: *A Short Primer on Six Sigma* Transfer Functions). User stories' priority profile is calculated by counting the number of data movements needed per user story to meet the IoT Needs' priority profile.

This profile is found at the bottom of the following transfer function (Figure 3-6) that computes functional effectiveness with these five data movements yields:

Figure 3-6. IoT Needs Coverage by Data Movements

IoT Needs Deployment Combinator		Goal Profile	Q001 Search Data	Q002 Answer Questions	Q003 Keep Data Safe	Achieved Profile	
y1	Extensible	0.53	3	1	1	0.50	
y2	Open	0.45		4		0.48	
y3	Reliable	0.68	2	2	3	0.70	
y4	Fast	0.22	1		1	0.18	
Solution Profile for User Stories:			0.55	0.68	0.49	Convergence Gap	
						0.06	

18 Total Effort Points
0.10 Convergence Range
0.20 Convergence Limit

The priority profile reflects the number of data movements needed in the software to cope with the user requirements expressed in user stories.

The test stories in turn are simple. Basically, the tests verify that data is kept safe and not altered when reading. Moreover, an invalid search string – whatever that means – is rejected and not used for searching the database. Missing data is shown as not available in the database, and repeatedly entering the same equation returns identical answers.

Figure 3-7: Test Stories with first Test Cases

	Test Story	Case 1	Test Data	Expected Response
A Prepare	A.1 Retrieve Responses	A.1.1	Search String; Valid	Return (known) answer
	A.2 Detect Missing Data	A.2.1	Search String; Valid; No Search Result	No response available
B Response	B.1 Validate Responses	B.1.1	Search String; Valid	Correct responses
	B.2 Data Stays Untouched	B.2.1	Query; Repeated	Return identical Answer

Instead of full test case assertions we use an abbreviated form that just indicated what test data should be specified here. Data can be specified as anything that matches a predicate such as $x < b$, or $a < x < b$. In view of section *1-2.2: A Representation for the World of Tests* care must be taken that in order to execute any such test, a mechanism must exist that selects an appropriate test data sample x; once more explaining why computer scientists must master intuitionistic mathematics, not traditional analysis. Every programmer knows how much can go wrong with such test data predicates that do not exactly specify how to pick an appropriate sample for executing the test.

The remaining test cases for two of the test stories are shown in Figure 3-8.

Figure 3-8: Test Stories with remaining three Test Cases

	Test Story	Case 2	Test Data	Expected Response	Case 3	Test Data	Expected Response
A Prepare	A.1 Retrieve Responses	A.1.2	Combined Query; Valid	Return (new) answer	A.1.3	Combined Query; Invalid	No response available
	A.2 Detect Missing Data						
B Response	B.1 Validate Responses	B.1.2	Search String; Invalid	Invalid search string			
	B.2 Data Stays Untouched						

The data movements executed for the first test case of the first test story A.1.1

A. 1.1: {Search String; Valid} → Return (known) answer

consists of the first three data movements:

Figure 3-9: Test Case A.1.1

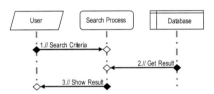

Thus, its test size is three. Moreover, the *User* device needs test stubs allowing him to get pairs of combined queries and known answers to execute this test case.

Completing the count yields the test coverage matrix (Figure 3-10):

Figure 3-10. Test Coverage for Simple Data Retrieval Application

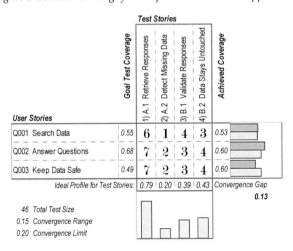

In general, every device object in a data movement map needs the ability to access test data and expected responses for executing tests. Some functional and data processes might need this as well, depending upon which test stories are defined. This is an additional task that developers must accomplish when preparing software for ART.

The test coverage transfer function in Figure 3-10 is defined by the number of data movements in a test story delivering user stories. Coverage is fine with a convergence gap of 0.13 in this transfer function, the total number of tested data movements per cell never exceeds seven. Total test size is 46, for a functional size of 5. Better convergence gaps are difficult to reach because of the small numbers.

3-2.2 CONNECTING IoT DEVICES TO THE DATABASE

We now add a sensor and an actuator to our IoT concert (Figure 3-11):

Figure 3-11: IoT Concert After Adding a Sensor and an Actuator

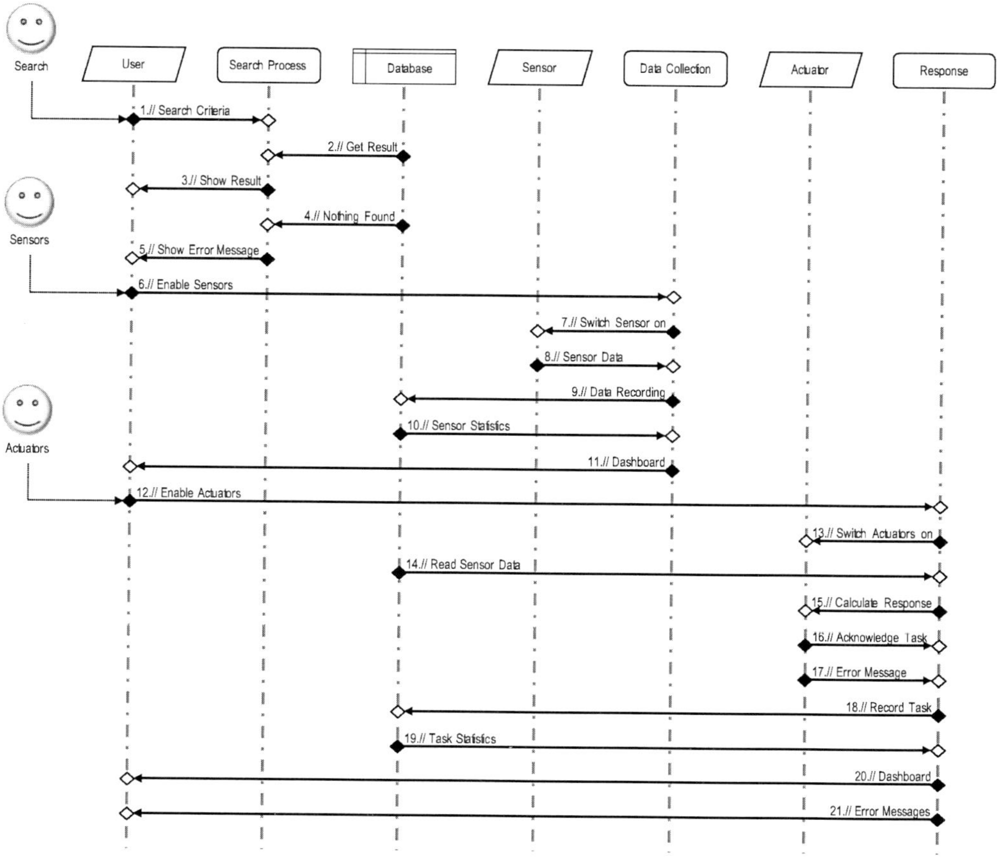

By adding one type of sensor and one type of actuator, the functional size almost triples and becomes 21 CFP. Security and safety risks increase with every data movement added to the IoT concert, as they can be misused or hacked, or cause unwanted and unsafe behavior.

3-2.2.1 ADDING MORE DATA MOVEMENTS

In practice, adding an IoT device goes with little or no programming. The additional devices come with software already prepared and use standard interfaces to connect with the database in our simple search module.

Nevertheless, there are a couple of new objects that require test stubs, making it obvious that ART is not something already there yet. Software suppliers need to cooperate to prepare their pieces for ART. In Figure 3-11, both the *Sensor* and the *Actuator* need such test stubs.

For the purpose of demonstrating ART, we keep the number of user stories and consequently of test stories, thus concentrating still on the same requirements while ignoring any additional requirements that could govern the use of sensors and actuators. Consequently, actuators and sensors will not be tested, as is probably realistic since we buy products ready for plug-in. If the application domain is rather safety-critical, such an assumption is potentially dangerous.

Figure 3-12: Functional Effectiveness after Adding an IoT Concert

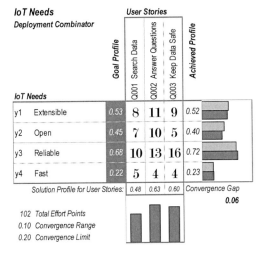

Functional effectiveness for the IoT concert is now expected to change (Figure 3-12), while the user stories and their profile remain. There are now many more data movements that impact user stories. Basically, these are the **Read** and **Writes** as defined by COSMIC to the database from both the functional processes that manage the sensor and the actuator. Note that the actuator also records the tasks it performs, adding more data than just sensor data to the database.

Since the user stories remain unchanged, the only interest is in verifying extensibility, openness, reliability, and access speed of the data already existing, or stored by the new sensor and the new actuator in the data base.

The *IoT Needs* deployment combinator for the full IoT data retrieval concert now takes more data movements into consideration, and consequently the user stories' profile changes (Figure 3-13). The goal profile for *IoT Needs* remains the same – not necessarily in all cases; however, no additional *IoT Needs* arise in this context with the full IoT concert, because it still does data retrieval, see Figure 3-13:

Figure 3-13. User Stories' Priority Profile for Full IoT Concert

	User Stories Topics	Weight	Profile
1)	Q001 Search Data	28%	0.48
2)	Q002 Answer Questions	37%	0.63
3)	Q003 Keep Data Safe	35%	0.60

3-2.2.2 EXTENDING TEST CASES

Functional size increases from 5 CFP (Figure 3-1) to 21 CFP (Figure 3-11) because of the added sensor and actuator and their respective functional processes for sensor data collection and for creating a response through the actuator. Also, user stories remain the same, although data now refers not to static but to dynamic data and the priority profile now changes towards higher importance for *Q003: Keep Data Safe*.

Figure 3-14: Extended Test Cases for the Full IoT Concert

	Test Story	Case 1	Test Data	Expected Response	Case 2	Test Data	Expected Response
A Prepare	A.1 Retrieve Responses	A.1.1	{Enter valid Search String}	Return (known) answer	A.1.2	{Combined Query; Valid}	Return (new) answer
	A.2 Detect Missing Data	A.2.1	{Search String; Valid; No Search Result}	No response available	A.2.2	{Sensor Off}	No data available
B Response	B.1 Validate Responses	B.1.1	{Search String; Valid}	Correct responses	B.1.2	{Search String; Invalid}	Invalid search string
	B.2 Data Stays Untouched	B.2.1	{Query; Repeated}	Return identical Answer	B.2.2	{Transmission Interference}	Data Rejected

Case 3	Test Data	Expected Response	Case 4	Test Data	Expected Response	Case 5	Test Data	Expected Response
A.1.3	{Combined Query; Invalid}	No response available	A.1.4	{Sensor Readings}	Retrieved in Database	A.1.5	{Transmission Error}	No Data available
A.2.3	{Sensor Off}	Dashboard Indication	A.2.4	{Actuator Off}	Dashboard Indication	A.2.5	{Invalid Actuator Data}	No Action
B.1.3	{Actuator Set}	Actuator does it						
B.2.3	{Transmission Interference}	Dashboard Indication	B.2.4	{Actuator Off}	Dashboard Indication			

Case 6	Test Data	Expected Response	Case 6	Test Data	Expected Response	Case 6	Test Data	Expected Response
A.1.6	{Actuator Enabled}	Dashboard Indication	A.1.7	{Actuator Off}	No Action	A.1.8	{Actuator Response}	Stored in Database
A.2.6	{Invalid Actuator Data}	Dashboard indication						

Test stories too remain the same but must cover additional data movements between devices, database, sensors, and actuators. Consequently, the IoT Needs profile (Figure 3-3) remains valid while the user stories' priority profile (Figure 3-5) changes after connecting the database to the IoT concert. Figure 3-5 transforms into Figure 3-13 with more focus on *Q003: Keep Data Safe.*

Consequently, test cases increase in number (Figure 3-14). For instance, to keep data safe (*Q003: Keep Data Safe*), data transmissions to sensors and actuators must be tested against loss of data, or data transmission interference, e.g., by hackers. This increases the test size but not the number of test stories.

Because of adding sensor and actuator, the number of test cases increases by all the new combinations of reading from and writing into the database. Additional test cases become necessary to test these assertions, such as test case A.1.4:

A. 1.4: {Sensor Readings} → Retrieved in Database

The corresponding test case uses the following data movements:

Figure 3-15: Test Case A.1.4

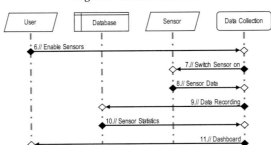

Figure 3-16. Test Coverage for Full IoT Concert

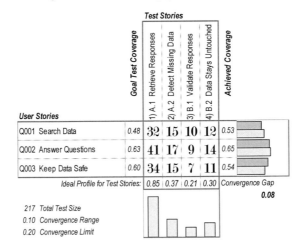

User Stories	Goal Test Coverage	1) A.1 Retrieve Responses	2) A.2 Detect Missing Data	3) B.1 Validate Responses	4) B.2 Data Stays Untouched	Achieved Coverage
Q001 Search Data	0.48	32	15	10	12	0.53
Q002 Answer Questions	0.63	41	17	9	14	0.65
Q003 Keep Data Safe	0.60	34	15	7	11	0.54
Ideal Profile for Test Stories:		0.85	0.37	0.21	0.30	Convergence Gap
						0.08

217 Total Test Size
0.10 Convergence Range
0.20 Convergence Limit

The resulting test coverage (Figure 3-16) remains like Figure 3-10 although test size increases considerably. This means that many more data movements are now under test; however, with the same test stories. The knowledge for testing the IoT is inherited from the original tests for the simple data retrieval test scenario.

Clearly, both transfer functions for both test coverages remain within a safe *Rule Set Radius*. Adding more types of IoT devices causes the cell counts grown in the test coverage matrix while the convergence gap remains within the rule set radius limits thanks to additional test cases. This is what combinatory logic predicts. Thus, the original data retrieval application test serves as a model for the full IoT test. Only one rule set has been applied so far: $(x_3 \rightarrow y)_3$, representing the transfer function for test coverage (Figure 3-16).

If the IoT concert covers more user stories, say j, then this becomes $(x_3 \rightarrow y)_j$; what in turn most likely requires i more test stories: $(x_i \rightarrow y)_j$. The importance of the original three test stories changes between the data retrieval application and the full IoT concert, like seen in Figure 3-5 and Figure 3-13.

Table 3-17. Test Priority Change when Adding Full IoT Concert

	Test Story	Data Retrieval		Full IoT Concert	
		Weight	Profile	Weight	Profile
A Prepare	A.1 Retrieve Responses	44%	0.79	49%	0.85
	A.2 Detect Missing Data	11%	0.20	22%	0.37
B Response	B.1 Validate Responses	21%	0.39	12%	0.21
	B.2 Data Stays Untouched	24%	0.43	17%	0.30

Main focus remained on *A.1: Retrieve Responses* but secondary changed from *B.2: Data Stays Untouched* to *A.2: Detect Missing Data*. This reflects the addition of tests that detects the failure of writing data from sensor or actuator into the database. This reflects the additional effort that is required to protect data movements between sensors and database from interferences, e.g., data loss or even privacy violations, reflecting the higher focus on *Q003: Keep Data Safe*,

Table 3-18 shows a comparison of test sizes between the original data retrieval application test, and the full IoT concert test.

Table 3-18. Data Retrieval Test Size vs. IoT Test Size

Data Retrieval		Full IoT Concert	
Test Size in CFP:	46	Test Size in CFP:	217
Test Intensity:	9.2	Test Intensity:	10.3
Defect Density:	40.0%	Defect Density:	19.0%
Data Movements Covered:	100%	Data Movements Covered:	100%

The key indicator for tests is the *Test Intensity*, the ratio between *Test Size* and *Functional Size*. *Defect Size* in turn is the percentage of defective data movements in the software. Size measurements follow the international standard ISO/IEC 19761

COSMIC (ISO/IEC 19761:2019, 2019). There are no limits for neither functional size nor test size.

3-2.3 AUTOMATED TEST CASE GENERATION

Thanks to the test priority goal profile, derived from the original IoT Needs and carried through user stories to test stories, it is possible to generate test cases automatically. The convergences gap serves as the heuristics which test cases to add to the test. The principle behind artificial intelligence are heuristic hash functions, i.e., metrics that tell which search branches to follow and which to avoid.

Because of the heuristics, artificial intelligence adds only test cases that pertain to the functionality of the implemented user stories, notwithstanding whether the IoT concert now features additional but untested functionality. The data retrieval approach does not cover additional requirements that might come with the IoT concert, such as in a smart house, whether window stores close when the sun is shining strong, or when it is necessary to avoid car collisions. The model extends, for instance, from the simple, well-controllable and well-tested application to something more sophisticated such as a smart home, or autonomous cars. Then, combinatory logic extends the test suite from the original model to the full-blown system.

3-3 CONCLUSIONS AND NEXT STEPS

Automated testing is a must for IoT systems, especially for autonomous cars. But automation is not enough. Autonomous testing means that new test cases are generated when software is updated or cloud services change. This requires a sound theory how to generate test cases and intelligence for selecting the relevant test cases for test execution.

The time for actual testing can be very small. For instance, in case of an encounter with another car from a different manufacturer that wants to connect and whose behavior is hardly predictable, testing time allowance might be reduced to a few seconds and therefore must be reduced to the minimum.

We demonstrated with an example how testing scenarios carry over from simple applications to complex IoT concerts, using the original test cases as testing patterns for automatically extending the test to the full IoT application. Using combinatory logic, testing scenarios designed for the original model carry over to its extended IoT implementation, and this is already an important saving, enabling safe IoT concertation.

Combinatory logic paves the way to testing complex IoT concerts and networked systems, based on the solid ground of existing testing experiences. The quality of testing can be maintained even after moving to automated testing. For testing the IoT, this approach offers significant savings; however, the full potential of combinatory logic in organizing knowledge is significantly greater.

Adding more "things" to that system requires additional testing that prove safety and security, other qualities, and functionalities of the expanded system. Such systems, serving as proof of concept, seem within easy reach for the currently available tools and can be used to study the legal basis for future, even more intelligent and autonomous things.

CHAPTER 4: TESTING PRIVACY PROTECTION AND SAFETY RISKS

Privacy protection has become a major concern since we noticed that Google always knows where we are – because of the location services switched on in our Smartphones. And because we find it so attractive to know where we are, to see which restaurants are open around us, what they offer, and investigate the shops' offers already before visiting them.

Maybe all this loss of privacy is not indispensable but who cares? On social media, we give even more insights into all aspects of our private life, and we know that a dozen characteristics are enough to match a person's identity even without the consent of people to disclose their names.

Unfortunately, privacy protection is more than just luxury. If software-intense products become popular, it is easy to use them for stealing relevant information, concerning money, property, or simply turn such products into a threat for your health or even life.

Privacy protection and safety risk assessment by Autonomous Real-time Testing (ART) is the foundation of digitalization. Digitalization cannot work without it. Or, what do you think will happen after the first incidence of the sort that your smartphone threatens you with causing an accident by misguiding your car? Unless you pay immediately some ransom fee? By bitcoin? Unfortunately, you downloaded a new, cool, app that tells your car's Advanced Driving Assistance System (ADAS) where to go...according your preferences, they said...

4-1 INTRODUCTION

While test intensity certainly is important, it is not a consumer metrics by itself. Consumers value more to know the degree of protection against perceived dangers. Among them, physical safety matters most when sitting in an autonomous vehicle, but privacy is another major concern. Not only is it sometimes not convenient if the public knows where the car was directed, but other aspects of privacy might be equally important. For instance, who overhears private conversations in a car? Who has access to the credit card used to refuel the car, or reload batteries? Some might be worried of hackers that might gain control over the car (Greenberg, 2015).

Privacy protection is not a new requirement. For centuries, privacy was easy to protect but hard to break because you had to personally overhear talking, not targeted at the public, or steal physical things such as letters or notepads. Nowadays, Amazon's Alexa can overhear you while you think you privately chat to friends and family, listening to music, or laptops can use their cameras watching you, and anyway, whatever you like, comment or disgust in newspapers and other social media is immediately known to almost everybody, be it the Russian secret service, the FBI, or Amazon and Google.

Nevertheless, you own the data that you produce and most of your listeners require permission to track you. Some services track you but anonymously; for instance, car drivers are traced and monitored by whatever navigation service they are using, not only for placing advertisement nearest to their location, but also to learn about traffic interruptions and jams.

While location is not so much a concern for most people, some people feel less at ease with the continual location tracking, be it when conducting secret visits for business talks or personal affairs, or simply when robbing a family home. Switching off your smartphone is a means of protecting your privacy; however, then you cannot use any of the features offered and since we all depend from our smartphones, you don't do it easily.

More serious is that hackers use personal data such as credit card numbers or passwords or both for stealing more tangible things such as money. Or they block entry into your well-protected IoT-controlled family home, asking you for ransom money before unlocking, eventually. Similar things can easily happen to your car, for instance by taking control over your *Advanced Driving Assistance System* (ADAS). If ever the dream of *Autonomous Driving* would come true, it could turn into a nightmare if the protection of privacy were insufficient.

4-2 CONSUMER METRICS

Figure 4-1: EU Energy Label

The EU has set a good example in the European Union energy label; see the Directive 2010/30/EU (European Commission, 2010) and Figure 4-1. A graphical representation is certainly better than simply presenting numbers. Thus, consumers can easily orient themselves.

If you want to get consumers to do tests, then you must think of something about how to present the results of such tests. James Watt had to explain how to compare the output of steam engines with the power of draft horses. The "Horsepower" is a unit of measurement for power – the rate at which work is done. It was later expanded to include the output power of other types of piston engines, as well as turbines, electric motors and other machinery.

The "Horsepower" unit of measurement became very popular, later, and still is, although it is at odds to the metric system.

We propose a graphical representation that uses similar colors and resembles the familiar FMEA diagram used in automotive, using two dimensions; see Figure 4-2:

- *Privacy Needs* – the level of protection needed, the worthiness of protection.
- *Privacy Protection* – the means used to protect data against theft or sniff.

Both dimensions use a zero-to-five scale, indicating the need for privacy protection and the means used to protect. While the privacy protection scale might be stable over time, the adopted means of privacy protection clearly are not and need consensus for acceptance. New protection schemes are easily fit into the zero-to-five scale.

The bubble marks where the system is placed in the grid in terms of privacy needs and privacy protection. The privacy index is the distance from the upper right corner – the worst case – to the bubble. Bubbles placed on the circles have the same index. The grid is skewed for accommodating bubbles that represent maximum protection even though they do not need it.

Figure 4-2: Proposal how to Assess Privacy Issues for Technical Systems

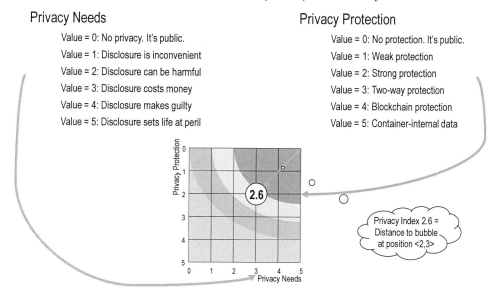

More than one bubble can be placed on the grid. This is useful if parts of the systems exhibit a different privacy behavior than others. The size of the bubble can then be used to indicate which one is predominant. If so, it is recommended to label the different bubbles indicating for what they stand for.

Privacy protection can be excellent if no privacy is needed because data is public. Public data does not need protection. It depends from the context. Container-protected data remains within a virtual machine and is not exposed to the environment. In view of the possibility of attacks to hardware – for instance *Spectre* and *Meltdown* – even container-internal data in containers that share one kernel is not entirely safe (Graz University of Technology, 2018). For consumer metrics, this limitation is acceptable.

Measuring privacy is basically the product of privacy value for the user times the degree of public exposure. If one of them is near zero, there is no privacy, or no privacy needed. Highest privacy protection is if there is data worth protecting, and protection is effective.

The formula for the privacy index is given in (4-1) on page 70 where *Needs* and *Protection* are the distances in the grid from the worst-case point, and thus must be counted inverse for the *Needs*. It is simply the Euclidian distance, somewhat distorted by allowing for the green bottom row. For a sample graphical representation, see Figure 4-3. The *Look & Act* application is explained later in section 4-3.1.

Figure 4-3. Privacy Needs vs. Privacy Protection for Look & Act

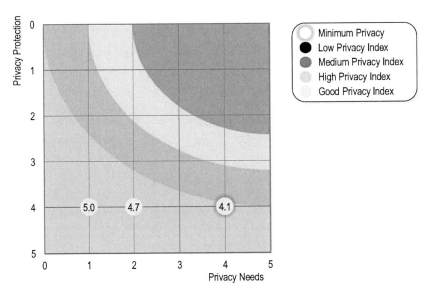

Much more elaborate schemes exist for characterizing privacy protection, distinguishing up to seven dimensions of protection, and for safety risk assessments, see e.g., Tilghman (Tilghman, et al., 2014) for warfare applications. While such specialized high-tech applications doubtless would benefit from autonomous real-time software testing, seven dimensions of protection indices seems far away from a privacy protection index representation acceptable to the public.

Consumer metrics do not replace sound technical assessments but help engineers identify weak points – at least those that impact consumers' perception. For a sound privacy protection and safety risk assessment, the traditional methods are still presumed; they remain indispensable. Moreover, some of the consumers' assessment criteria cannot be answered without knowing the technical background. For instance, container protection depends from the implementation details and is not replaceable by consumer metrics.

We recommend limiting the number of bubbles shown to consumers. For instance, more bubbles than shown in in Figure 4-3 are not helpful. One, or two, plus the minimum privacy, or maximum risk, bubbles are enough. The tools support limiting the number of bubbles. However, a general recommendation cannot yet be given.

4-2.1   ACCEPTABILITY OF CONSUMER METRIC FOR SAFETY & PRIVACY

There are important obstacles to overcome for such consumer metrics. The first one is that suppliers of autonomous vehicles are not very eager of getting measured by anyone, and if performing measurements, to keep results under disclosure. Another one is that customer organizations, forget lawmakers, have not yet fully understood the impact of digitalization and, in turn, of autonomous vehicles on the society.

Nevertheless, users of laptops and smartphones would already today welcome such indicators after downloading new apps or an operating system update, or after new attacks on their privacy have been publicly communicated. Whoever comes first proposing such consumer metrics might gain a significant competitive advantage, forcing the competition to follow up.

An open question remains whether assessing privacy and safety on data movements alone is good enough for all domains. While this choice has obvious advantages for cloud systems, the *Internet of Things* (IoT) (Fehlmann & Kranich, 2017), and embedded software in autonomous vehicles, it is unclear whether it also suffices for mobile applications, or traditional web applications and commercial software.

4-2.2   THE PROPOSED MEASUREMENT PROCESS

More difficult obstacle is that the proposed measurement process uses models for large and complex software systems that are far from widespread practice. While the IFPUG model is popular for early cost estimation, and the COSMIC model is used for estimation of memory load prediction in automotive (Soubra, et al., 2015), it is generally difficult to get an accurate model after completion of the software, or for any software in operation.

For institutionalizing consumer metrics for software, the software deployment process – aka DevOps toolchain – needs being enhanced to provide such models. Luckily, at least for the ISO/IEC 19761 COSMIC model, automated model creation tools, suitable for model-based testing as well as for consumer metrics, are available (Soubra, et al., 2014). Microservice architecture, based on Kubernetes (The Kubernetes Authors, 2018), almost instantly transform into a COSMIC model. Where automated measurement tools are not available yet, models can still be created manually, as for predicting cost; however, this is costly, and keeping those models updated for new releases is challenging.

4-2.3 PRIVACY WITH AN ADVANCED DRIVING ASSISTANT SYSTEMS

The sample ADAS service (Figure 4-4) demonstrates the principles. This is a *Car Driving Function* based on a visual recognition system (Camera driven by a *Sensor Bus*) and a *Neural Network Engine* interpreting images. A *Lidar* – a device that measures distances with a pulsed laser light – delivers distances and allows the neural network engine to assess the safety risks that originate from the object on the image analyzed. Sequences of images serve for determining the objects movements and direction.

Figure 4-4: Look & Act in ADAS

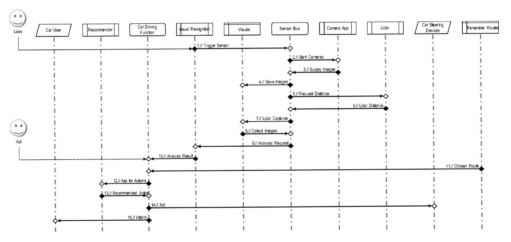

The *Car Driving Function* asks the *Recommender* for advice and *Acts* in accordance with the selected route that the navigation system stored in the *Remember Routes* persistent database. This is a simplified ADAS for instructional purpose only; it possibly can power a model car. But it is a model car equipped with camera, Lidar, and sensors for slippery roads. And it uses a Navigator service to find a route. However, we outsourced both the recommender and the *Visual Recognition System* (VRS) which do most of the work. Both services are likely implemented as neural network engines. Nevertheless, for a real-world ADAS, our sample application lacks redundancy.

Privacy is best measured by looking at the data movements between objects, under the assumption that the application objects do no other data movements than those listed in the model. Compliance to the ISO/IEC standard 14143 ensures exactly this (ISO/IEC 14143-1:2007, 2007). Then privacy protection is measurable by the protection level of the data movements between those objects.

These data movements can be open to the public, encrypted, or secured by two-way authentication scheme, by blockchain, or be transported on physically isolated and protected cables. Data groups moved within a container, or processor, are among the latter. Data protection within containers cannot be taken for granted but can be assured with reasonable effort. For the technology behind such protection, see e.g., Staimer (Staimer, 2015).

Protection methods in turn are implementation dependent – and the labels chosen arbitrary. Protection technology will change, and encryption might be appropriate in many cases to protect data movements when transporting data through Internet, but there exist many industrial bus systems that require different categories with different labels. Also, encryption is not the only way protecting data against hackers. Encryption is best against man-in-the-middle attacks, but many more attack vectors exist and many other effective protection schemas. Again, only the level matters.

Table 4-5: Privacy Assessment Categories

Privacy Needs		**Privacy Protection**	
Value = 0:	No privacy. It's public.	Value = 0:	No protection. It's public.
Value = 1:	Disclosure is inconvenient	Value = 1:	Weak encryption
Value = 2:	Disclosure can be harmful	Value = 2:	Strong encryption
Value = 3:	Disclosure costs money	Value = 3:	Two-way encryption
Value = 4:	Disclosure makes guilty	Value = 4:	Blockchain protected
Value = 5:	Disclosure puts life at peril	Value = 5:	Container-internal data

These categories can be used in a table for recording the assessment.

The Table 4-5 above shows five categories of privacy needs on the left and five degrees of privacy protection methods on the right. Privacy needs can be directly assigned to data groups in COSMIC; this is a model property. The labels chosen are unimportant, the level matters.

For a graphical representation, we propose a square representation. This also explains why we consider two dimensions only; for consumer metrics, this is already challenging.

Distance of the bubbles in the grid (Figure 4-3) is measured from the starting point (0,0). The Privacy Index is in range 0 – 5. Five (5) is the index for maximum privacy; Zero (0) privacy means public data; no privacy granted, or no privacy needed.

The Privacy Index should provide equal length for equal protection; thus, Euclidean distance yields the following, square root of sum of squares, formula:

$$Privacy\ Index = Max\left(5, \sqrt{\left((5-Needs)*{}^5\!/_5\right)^2 + \left(Protection*{}^6\!/_5\right)^2}\right) \quad (4\text{-}1)$$

The *Needs* coefficient must be inversed by 5 because the bubble distances are calculated from the upper right edge of the graph area. The maximum function in equation (4-1) ensures that the index is bounded by a maximum of five. In Figure 4-3, the size of the bubbles indicates how many data movements yield that index. The minimum privacy – here 4.1 – is highlighted.

Stretching the *Protection* by ${}^6\!/_5$ has the effect that if no privacy is required, the privacy index remains high (upper left square in Figure 4-3).

The size of the bubbles corresponds to the number of data movements that lie within this privacy index range. The graphical representation in Figure 4-3 is intuitive because distance from the upper right square conforms to the level of privacy, which is best at the down-right square. It has some resemblance to the FMEA Criticality Matrix of the *German Verband der Automobilindustrie* (VDA, 2008, p. 64) and thus is also acceptable to automotive security engineers.

4-2.4   SAFETY RISK

Safety risks are less difficult to represent. According classical risk management theory (ISO 31000:2018, 2018), risks can be assessed by

- Identifying the risk catalogue
- Classify impact, usually on a scale 0 – 5
- Assigning the probability of risk incurrence

For identifying safety risks in road vehicles, the series of international standards ISO 26262 (ISO 26262-1, 2011), provide guidance. Recently, the new SOTIF[1] version of the ISO/IEC 26262 has been released. These standards can be used for assessing risks of critical parts; not only mechanical, but also data movements moving critical data groups.

---

[1] SOTIF = Safety of the Intended Functionality

Since we avoid fake assessment precision, we use the same scale 0…5 for probability as well, thus only allowing for 0%, 20%, 40% risk probability. Moreover, probability is something difficult to find in software; we use frequency instead, namely the frequency of executing a certain data movement. Frequency is an implementation characteristic and cannot be assessed uniquely in the model. The risk of *Safety Impact* on the other hand depends from the content of the data group and is a model property, like the privacy needs in privacy protection assessment.

The safety risk graphical representation for consumers looks as follows:

Figure 4-6. Safety Risk Exposure for Look & Act

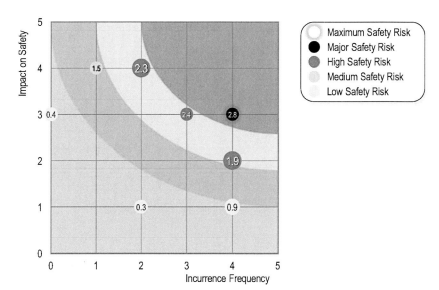

The *Safety Index* is calculated as follows:

$$Safety\ Index =$$

$$Min\left(5, \sqrt{\left((5 - Probability) * \frac{5}{5}\right)^2 + \left((5 - Impact) * \frac{6}{5}\right)^2}\right) \qquad (4\text{-}2)$$

For the graphical representation, we propose the equation (4-2), which looks similar to (4-1), again using Euclidian length for the positioning of the bubbles. Because distance in the risk grid is measured starting from the ⟨5,5⟩–Point, both grid indices will be mirrored at the grid size value 5, and colors should remain the same for the consumer.

Table 4-7: Safety Assessment Categories

Incurrence Frequency		Safety Impact	
Value = 0 (0%):	No risk. It's safe.	Value = 0:	None
Value = 1 (20%):	Seldom	Value = 1:	Low
Value = 2 (40%):	Sometimes	Value = 2:	Little
Value = 3 (60%):	Medium	Value = 3:	Medium
Value = 4 (80%):	Often	Value = 4:	Quite
Value = 5 (100%):	Very frequent	Value = 5:	High

The safety risk graph yields different information, showing that the various data groups in *Look & Act* move data of unequal impact on safety. The most impact (*Maximum Risk Index 2.8*) originates from data movement *10) Analysis Result*; by lack of redundancy – or lack of check by another "intelligent" module – its frequency is *1: Seldom* and its impact *4: Quite*. Reducing impact to *2: Little* could be achieved with adding a cross-check against serious impact, or using two independent *Recommenders* that agree on actions.

4-2.5 PERFORMING THE ASSESSMENTS FOR PRIVACY & SAFETY

The assessment is part of the COSMIC model and can be recorded directly in the table for data movements. *Privacy Needs* are represented by the effects of privacy disclosure.

Figure 4-8: Assessment of Look & Act Data Movements

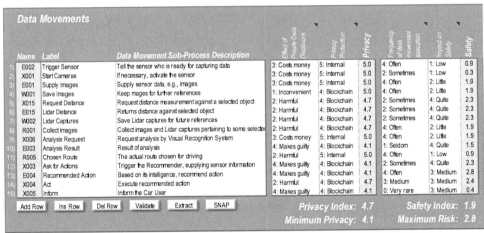

Figure 4-3 and Figure 4-6 have enough similarities to help consumers understanding the meaning of both indices, such that they can look at both representations together and get a correct impression. Figure 4-8 shows the data represented by the bubbles.

The question is how the *Privacy Index* (4-1) and the *Safety Index* (4-2) should combine for all data movements assessed. We prefer the *Median* against *Average* because the median is less subject to the effect of outliers. However, one outlier is always import, namely minimum privacy and maximum risk. They mark the weakest points in the system, and consequently the likeliest violation locations. Nevertheless, outliers are bad representative for the whole system.

4-3 ART FOR ADAS

The full ADAS application for our model car consists of four more parts:

- *Find Route*, e.g. by help of a navigation system, or according car user's preference.
- *Locate*, compare current location with actual route.
- *Check Route*, used to compare different possible routes in terms of traffic, weather, any other obstacles or fitting car user's preferences.
- *Amend Route*, after conditions changed under way it can become necessary to propose another route.

4-3.1 ADAS FUNCTIONALITY

Finding a route is usually based on some Navigator service (see section 2-2.7) that can propose a route between current location and some known destinations.

Figure 4-9: Find Route using Navigator and GPS Services

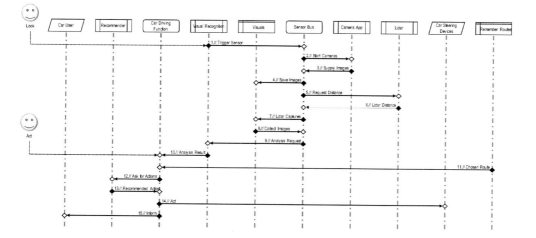

Thus, it is necessary to keep the car user informed in case no route is selected.

Location service is used to show the user where the car is driving:

Figure 4-10: Location using GPS

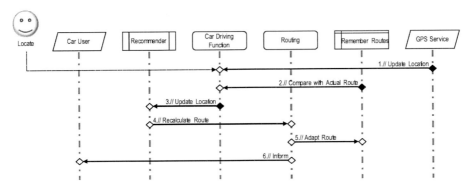

Checking the route involves rejecting a proposed route and selecting another one – or none if none is left. If none is left, the ADAS eventually cannot continue and manual driving is necessary. Because the Look & Act part requires knowing where to go, the ADAS is significantly less useful without a route selected. The complete ADAS is shown in Figure 4-13 on page 76; results of joining Figure 4-4 with Figure 4-9, Figure 4-10, Figure 4-11, and Figure 4-12.

The car driver may want to select another route, or the Navigator offers a selection of possible routes:

Figure 4-11: Approve or Modify Route

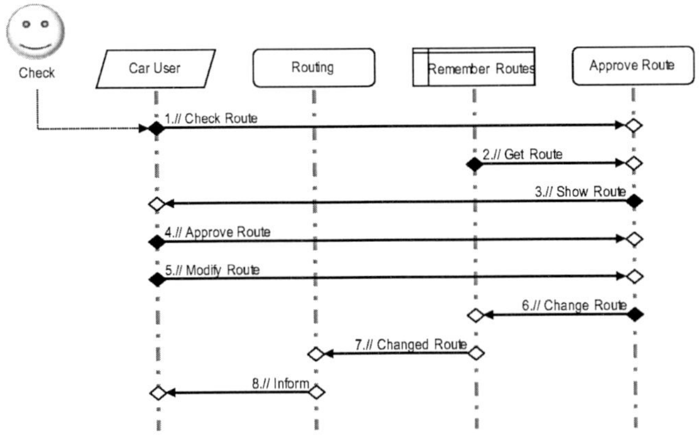

If a problem occurs with the selected route while driving, it can become necessary to amend the chosen route. We assume the Navigator service is capable of alerting in case of any change on the chosen route – which includes that the Navigator knows about the chosen route, eventually violating privacy of location.

The car driver is still entitled to choose yet another route, using Figure 4-11. The data movement *1.// Routing Alert* in Figure 4-12 proposes another route or amend at least the driving time prediction.

Figure 4-12: Alert on Chosen Route

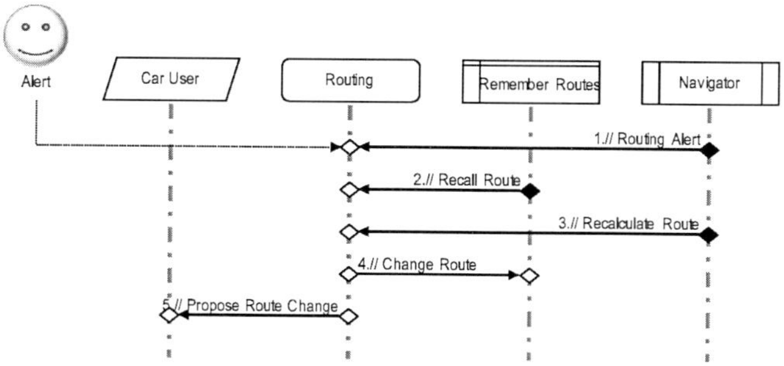

The full data movement map in Figure 4-13 on page 76 is the concatenation of these five parts.

4-3.2   TESTING THE ADAS

Now, in order to test all these services with regard to the assessed privacy protection and presumed safety risk exposure, one has to provide an *Automated Real-time Testing* (ART) application providing the necessary tests, such as verifying the encryption level per data movement as stipulated, and data group content according the assumption done in Figure 4-8. Note that the *Navigator* app provides not only routes but also driving conditions; part of the data group moved by the data movement *Routing Alert*.

This piece of software first prepares the setting – collecting car specifics, test cases, extending them – then executes testing first the neural network engine, then the recommender, finally the Lidar and the camera.

The testing software resides local, on the car, but the test data originate from a repository called *Testing Cloud* common to all cars undergoing the same tests. Test cases originate there, and the Testing AI engine also works on this cloud service. The ADAS of the car could upload images taken for adding those to the testing cloud; however, this is neither reflected in the part of the ADAS shown before, nor in Figure 4-14. Only test results are recorded in the testing cloud, upon approval by the car user, the owner of the test results.

Figure 4-14 on page 77 shows the data movement map for *Automated Real-time Testing (ART) for some Model of an Advanced Driving Assistant System (ADAS)*.

Figure 4-13: The Complete ADAS Model

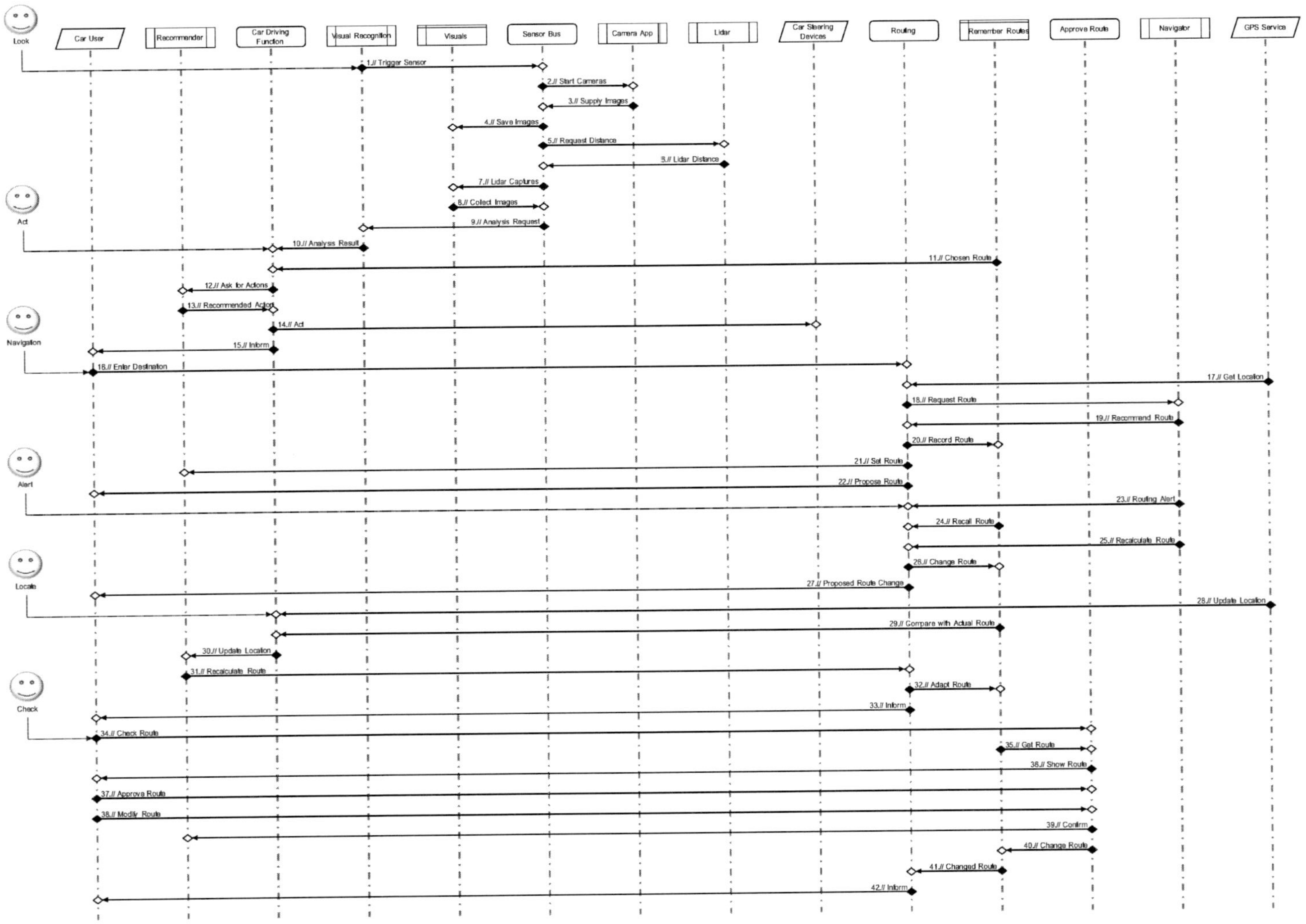

76

Figure 4-14: Automated Real-time Testing (ART) for some Model of an Advanced Driving Assistant System (ADAS)

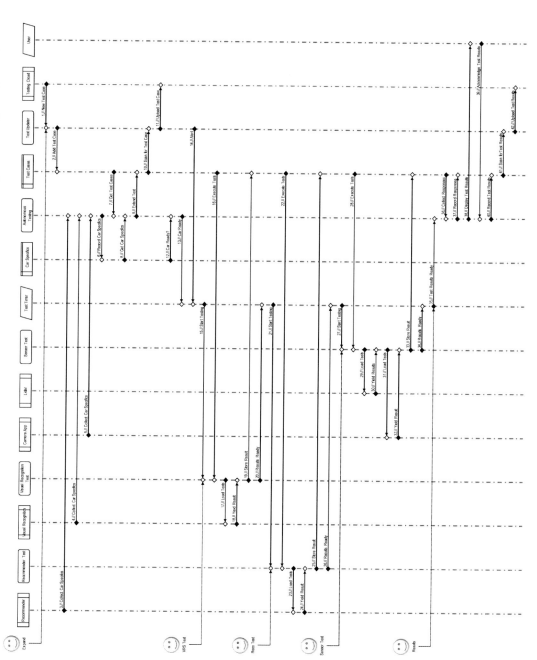

Figure 4-14 consists of test preparation, execution of tests for the *Neural Network*, the *Recommender*, and the *Visual Recognition Systems* including the *Lidar*, plus a test result recording and test result presentation for the tester testing the ADAS. It represents an application by itself, with user stories and the need for testing. However, since the main concern is getting the right kind of test cases that can be executed automatically, we keep the focus on testing ADAS (Figure 4-13).

4-3.3   THE CAR USERS' NEEDS

Using the AHP, we identify the following major values for users of the ADAS:

Figure 4-15: Car Users' Needs

	Car Users' Needs Topics	Attributes			Weight	Profile	
Y.a Drive Fast	y1 Agile Driving	Arrive safe	Do not block other traffic	Have fun	16%	0.34	
	y2 Smooth Driving	Drive predictibly	Do not break unnecessarily		14%	0.30	
	y3 Arrive in Time	Arrive predictibly	Avoid obstacles		23%	0.50	
Y.b Drive Safe	y4 Avoid Incidences	Drive foresightful	Know what's ahead	Know my way	27%	0.59	
	y5 No Surprises	Communicate	Never surprise anybody	Give signs	21%	0.45	

(AHP Priorities: Weight / Profile)

The AHP process is used to analyze these needs and produce a profile for its relative importance. The profile for the car users' needs is based on the following pairwise comparison, shown in Figure 4-16. This is again a basic AHP:

Figure 4-16: AHP for ADAS

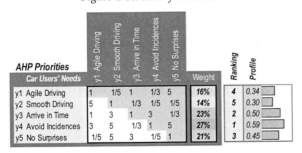

AHP Priorities Car Users' Needs	y1 Agile Driving	y2 Smooth Driving	y3 Arrive in Time	y4 Avoid Incidences	y5 No Surprises	Weight	Ranking	Profile
y1 Agile Driving	1	1/5	1	1/3	5	16%	4	0.34
y2 Smooth Driving	5	1	1/3	1/5	1/5	14%	5	0.30
y3 Arrive in Time	1	3	1	3	1/3	23%	2	0.50
y4 Avoid Incidences	3	5	1/3	1	5	27%	1	0.59
y5 No Surprises	1/5	5	3	1/5	1	21%	3	0.45

The needs of human drivers in today's traffic might be individually quite different; however, in view of an ADAS, characteristics linked to safety and avoidance of disturbance are dominant. You use an ADAS because you need something that helps through dense urban traffic, avoids jams and incidences, and makes driving experience smoother.

An ADAS is less suited for people who drive cars just for fun. They eventually turn it off. Their needs are not investigated by that AHP; an AHP for such people likely would produce a different car users' needs profile.

The data movements are those of the joint ADAS data movement map Figure 4-13. The user stories for ADAS are summarized in Table 4-17:

Table 4-17: ADAS User Stories

Label	As a ...	I want to ...	Such that ...	So that ...
Populated Area	Car User	let my car reduce speed	my car can safely stop	my car is not causing delays by an incidence
Obstacle	Car User	let my car avoid obstacles	my car can drive around	my car is not stopping unnecessarily
Know my Way	Car User	let my car take appropriate routes	my car avoids blocked routes and traffic jams	I know when I'll arrive
Amend my Way	Car User	optimize my route when needed	no incidence blocks my way	I still can predict when I'll arrive
Check my Way	Car User	approve or disapprove the car's choice for routing	I can take my preferred route	I feel in control
Able to Stop	Car User	have my car break soon enough	it can avoid dangerous situations	It recognizes obstacles ahead
Check my Way	Car User	approve or disapprove the car's choice for routing	I can take my preferred route	I feel in control

The user stories remain on a high epic level without specifying the details how the ADAS should behave in specific cases. With these user stories, the functional effectiveness matrix yields a satisfying rationale for the user stories (Figure 4-18):

Figure 4-18: Functional Effectiveness for ADAS

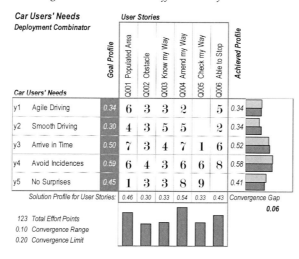

It means that the data movement map implements the user stories completely and without any wrong focus.

As before, the functional effectiveness transfer function maps the user stories onto the car users' needs by counting how many data movements contribute to the user stories. This yields the cause-effect relation between functionality and requirements; also, it assigns data movements to at least one user story.

4-3.4   THE TEST STORIES

The test stories tell more about the details how to implement ADAS functionality; see the following Table 4-19:

Table 4-19: Test Cases for ADAS

	Test Story		Case 1	Test Data	Expected Response	Case 2	Test Data	Expected Response
1)	A People Around	A.1 People around	A.1.1	{Playball; Populated Area}	Get ready to break	A.1.2	{Person; Moving; Towards street}	Stop before collision!
2)	B Obstacle	B.1 Obstacle ahead	B.1.1	{Obstacle ahead!}	Stop before collision!	B.1.2	{Obstacle; At roadside}	Drive around
3)	C Know my way	C.1 Get route	C.1.1	{Valid destination}	Select best route	C.1.2	{Invalid destination}	Select route home
4)		C.2 Change route	C.2.1	{Alert; Alternative available}	Propose new route	C.2.2	{Alert; No alternative available}	No better route available
5)		C.3 Update position	C.3.1	{Current position}	Recalcuate arrival time	C.3.2	{Route; Change}	Recalcuate arrival time
6)	D Choose way	D.1 Approval	D.1.1	{Route; Approval}	Confirm this route	D.1.2	{Route; Reject}	Propose another one
7)	E Arrival	E.1 Arrival time	E.1.1	{}	Show expected arrival time	E.1.2	{New conditions ≠ Route conditions}	Change expected arrival time
8)		E.2 Learnings	E.2.1	{Route; Fast}	Prefer them	E.2.2	{Route; Slow}	Avoid them
9)	F Stop	F.1 Keep under control	F.1.1	{}	Car can stop within sensor's reach	F.1.2	{Route conditions = bad!}	Lower speed
10)		F.2 Brake action	F.2.1	{Dry road condition}	Short braking distance	F.2.2	{Route conditions = wet}	Medium braking distance
11)		F.3 Avoid stops	F.3.1	{Under all conditions}	Listen to actual road condition	F.3.2	{Route; Traffic jam}	Try another route

	Case 3	Test Data	Expected Response	Case 3	Test Data	Expected Response	Case 3	Test Data	Expected Response
1)	A.1.3	{Person; Looking; At traffic}	Lower speed	A.1.4	{Person; Motionless}	Go ahead			
2)	B.1.3	{Obstacle; Light}	Drive around	B.1.4	{Route; Obstacle}	Change route			
3)	C.1.3	{Location = Home}	"Destination reached"	C.1.4	{Route}	Show risks	C.1.5	{Location}	Show position
4)	C.2.3	{Route; Modification}	Show risks						
5)	C.3.3	{Null GPS}	Continue current route	C.3.4	{Route; Changed; Approved}	Change current route	C.3.5	{Location}	Show position
6)	D.1.3	{}	Choose proposed route	D.1.4	{Route; No alternatives}	Choose proposed route			
7)	E.1.3	{Changed route}	New arrival time	E.1.4	{Route; New alert}	New arrival time			
8)	E.2.3	{Route conditions}	Adapt speed						
9)	F.1.3	{Route weather = bad!}	Reduce speed	F.1.4	{Route; Rain}	Reduce speed			
10)	F.2.3	{Slippery road}	Long braking distance	F.2.4	{Speed = low}	Short braking distance	F.2.5	{Speed = medium}	Medium braking distance
11)	F.3.3	{Red light ahead}	Lower speed	F.3.4	{Route; modification}	Show risks			

Read the test cases in Table 4-19 with an arrow → between test data and expected response. There are three more test cases for test story *10) F.2: Brake action:*

- F.2.6: {Speed = high} → Long braking distance
- F.2.7: {Must brake; curve} → Normal braking distance
- F.2.8: {Must brake; Descent} → Normal braking distance

Thus, for test story *10) F.2: Brake action* we have a maximum of eight test cases, where the other test stories only have five test cases or less, according Table 4-19.

This yields the following test coverage:

Figure 4-20: Initial Test Coverage

User Stories	Goal Test Coverage	1) A.1 People around	2) B.1 Obstacle ahead	3) C.1 Get route	4) C.2 Change route	5) C.3 Update position	6) D.1 Approval	7) E.1 Arrival time	8) E.2 Learnings	9) F.1 Keep under control	10) F.2 Brake action	11) F.3 Avoid stops	Achieved Coverage
Q001 Populated Area	0.46	25	22	9	7	11	9	10	8	12		14	0.42
Q002 Obstacle	0.30	10	15	13	5	15	7	11	9	13	16	10	0.36
Q003 Know my Way	0.33	2	5	17	6	15	12	9	6	7	9	9	0.27
Q004 Amend my Way	0.54	24	19	14	19	21	9	25	9	17	15	21	0.59
Q005 Check my Way	0.33	16	13	6	5	7	23	12	8			20	0.35
Q006 Able to Stop	0.43	26	25	5	2	10	4	6	8	10	8	13	0.39
Ideal Profile for Test Stories:		0.44	0.41	0.25	0.20	0.32	0.24	0.32	0.19	0.25	0.20	0.36	Convergence Gap

768 Total Test Size
0.15 Convergence Range
0.20 Convergence Limit

Convergence Gap 0.11

With a convergence gap of 0.11 we are within convergence range – set a bit wider than in usual transfer functions to allow for this small numbers of user and test stories.

4-3.5 EXTENDING TEST CASES

Extending test cases within the same test stories yields more reliable results, and a higher test intensity. In this example, extension works in two stages:

- Adding test cases that refer to bad weather forecast. If the *Navigator* reports rain on the route, driving speed and arrival forecast must be adapted.
- Even more test cases are added after the *Navigator* reports stormy weather causing eventually a change to the chosen route.

ART detects these new test cases because the data group received from the *Navigator* contains a weather forecast, as part of the route description; compare with Figure 4-12 on page 75. New test cases are created starting from the existing ones, by variation of test data, considering other all data received from data movements. Obviously, weather forecast changes the driving time prediction. Among the many test cases that can be created, ART keeps the convergence gap within limits, using this as selection process. The following matrices in Figure 4-21 and Figure 4-22 show the results after each of the two steps outlined above:

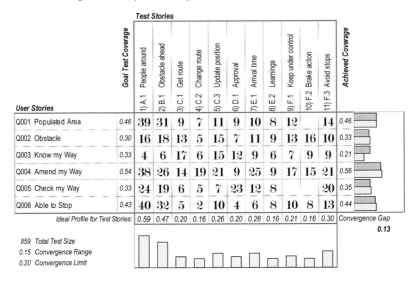

Figure 4-21: After Adding Bad Weather Forecast Test Cases

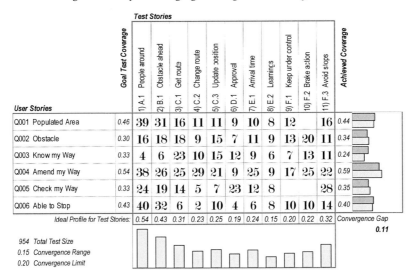

Figure 4-22: After Changing Routing due to Stormy Weather

Total test size is growing, and convergence gap is stable, or shrinking. The additional test cases improve reliability and accuracy. ART finds such extensions by scanning data groups of the data movements involved. Since the chosen route is not fix but changes on receiving an *Alert* from the *Navigator*, the VRS learns that conditions such as rainy and stormy weather can exist.

ART detects these new test cases because the data group received from the *Navigator* contains a weather forecast, as part of the route description. Obviously, weather forecast changes the driving time prediction.

4-3.6 SUMMARY VIEW

The summary view (Figure 4-23) on the original and the two extended test suites reveals, as expected, that test size and intensity increased.

Figure 4-23: Initial Test Suite, and two Extensions

Total CFP: 39	Test Size in CFP: 768
	Test Intensity: 19.7
Defects Found in Total: 0	Defect Density: 0.0%
Defects Pending for Removal: 0	Data Movements Covered: 100%

Total CFP: 39	Test Size in CFP: 859
	Test Intensity: 22.0
Defects Found in Total: 0	Defect Density: 0.0%
Defects Pending for Removal: 0	Data Movements Covered: 100%

Total CFP: 39	Test Size in CFP: 954
	Test Intensity: 24.5
Defects Found in Total: 0	Defect Density: 0.0%
Defects Pending for Removal: 0	Data Movements Covered: 100%

Functional size remained stable: CFP 39, while increasing test size also increased test intensity. Contrary to the IoT case, the functional size of the model ADAS remains the same.

Thus, improving testing is always possible by simply extending the test cases by similar ones, provided test coverage keeps the convergence gap narrow enough. ART provides value without increasing functional size. In this example, it was enough to trace back data movements that could contribute data to tests. Thus, the data movement map is paramount for automatic test case generation.

For testers, it suffices to provide an initial test suite (Table 4-19: *Test Cases for ADAS*). The rest is left to automatisms. You can increase test intensity as much as you like. More tests certainly increase opportunities for detecting defects that can be removed. Thanks to the test coverage transfer function and its convergence gap, those additional tests remain relevant. Moreover, since tests are generated randomly, there is no bias blocking certain test cases, although extending test cases along some application cases such as weather or route change might allow for targeted test extensions.

4-4 Conclusion

Testing Privacy and Safety is an ongoing task, that not only needs continual repeat but also extension in scope. What once was appropriate is within short time obsolete. Consumers have the right and the duty of keeping themselves informed about the actual status, and ART is delivering such updated and actual notification.

In the next chapter, we take a deeper look in how ART generate new relevant test cases within given test stories. In the end, ART uses methods from AI, and uses them to test AI. In some sense, ART applies the design ideas behind AI to the field of software testing.

CHAPTER 5: ARTIFICIAL INTELLIGENCE FOR TESTING

Artificial Intelligence for Testing provides test cases for extending test suites. The intelligence relies on finding variations of given test cases for a test story and selecting the right ones from these variations.

Using the data movement maps as a guide, generating new test cases can be accomplished by extending existing test cases by similar ones. The data group yields the relevant information in which direction to extend.

Such a process can be conducted with no limit. Nevertheless, for Autonomous Real-time Testing (ART), we also have the term "Real-time", and this means that we must be able to make selection small enough to fit into some available time allowance. This requires having some limiting function telling the AI robot when it is done.

AI for testing is expected to look at the software and to add test cases that prove the software's ability to achieve certain goals. To do this, goals of testing must be known, and the AI robot must be able to judge whether a test response is correct or not. The latter can be achieved by learning but also requires some understanding for the domain addressed by the software. For example, if the software drives a vehicle, a model must exist that allows the robot to decide whether an action proposed by the software under test is appropriate to achieve its goals.

5-1 WHAT IS THE GOAL OF TESTING?

As we have already seen, there is no automated testing without knowing the goals of testing. The goals must be available as a profile, clarifying priorities among the functionalities defined by user stories, or other means of expressing *Functional User Requirements* (FUR). Profiles are normalized as ever; see section 2-3 on page 28.

A *Topic* is something characterizing customer needs that the software under test shall deliver. It can be anything that is in use when talking about software, especially user requirements, business drivers, or business values. From this, a profile for the user stories can be derived using a transfer function. This derived profile is the goal of testing.

Let $y = \langle y_1, y_2, ..., y_n \rangle$ be a vector in the n-dimensional vector space of topics. We recall that the vector y is a *Profile*, if the equation (5-1) holds:

$$\|y\| = \sum_{j=0}^{n} y_j^2 = 1 \qquad (5\text{-}1)$$

As before, the double-bar $\|...\|$ indicates the *Euclidean Norm* for vectors. Any vector $x \neq 0$ can become a profile by dividing it through its length $x/\|x\|$.

Being Euclidian vectors, profiles can be compared. Also, profiles can be added or subtracted; however, then they lose the property of having length one unless you recalibrate the resulting vector on length one.

Assume two profiles $y = \langle y_1, y_2, ..., y_n \rangle$ and $z = \langle z_1, z_2, ..., z_n \rangle$, then its difference is:

$$y - z = \langle y_1 - z_1, y_2 - z_2, ..., y_n - z_n \rangle \qquad (5\text{-}2)$$

The difference is not necessarily a profile; however, equation (5-3) makes another profile out of the difference, provided the difference is not equal to zero. This profile points into the same direction as the difference vector but with a length of one:

$$\frac{y - z}{\|y - z\|} = \frac{\langle y_1 - z_1, y_2 - z_2, ..., y_n - z_n \rangle}{\sum_{j=0}^{n}(y_j - z_j)^2} \qquad (5\text{-}3)$$

The ability to compare profiles is the key to automated testing. Provided you have a goal profile, you can compare this goal to what you are planning to test. This comparison allows selecting test cases such that test effort remains limited, but the goal of testing is achieved within acceptable limits.

5-1.1 TRANSFER FUNCTIONS FOR TEST COVERAGE

The transfer function that defines the test stories needed to test a certain user story profile is called *Test Coverage*. Test coverage has a convergence gap that tells how well coverage is with regards to user stories. Since real-world user stories for software count for a few hundred rather than the half dozen shown with this book, test stories have similar dimensions.

However, since transfer functions can be computed quite effectively nowadays, this is not so much a concern. The test coverage matrix is automatically filled as soon as the functional effectiveness transfer function is established. The functional effectiveness matrix links data movements to certain requirements. However, since functional effectiveness has a convergence gap, the data movements' assessment can be validated.

5-1.2   What means Test Coverage?

Test stories and user stories complement each other. While the user stories explain what must be achieved, test stories often specify how this must be achieved. Thus, the test coverage matrix is a traditional QFD interrelationship matrix, matching the "how" to the "what" as explained in the respective ISO/IEC 16355 standard (ISO 16355-1:2015, 2015). If the convergence gap is small, it means that the test stories "implement" the user stories good enough. Or, in other words, the test stories test what the user stories require but nothing more.

It also means that nothing else than the user stories can be taken for granted. Properties not mentioned in user stories might hold or not; they remain untested.

The transfer function constitutes of the test sizes per user story. Each cell in the QFD matrix reflects the number of data movements executed by some test case in a test story that pertains to some user story. The functional effectiveness matrix is decisive for that. Because the assignment of data movements to user stories is sort of arbitrary, test coverage depends from which data movements are considered important or supportive for certain user stories.

5-2 Generating New Test Cases

Artificial Intelligence (AI) is not a well-defined notion. According TechTarget (Rouse, et al., 2018), AI is the simulation of human intelligence processes by machines, especially computer systems. These processes include learning (the acquisition of information and rules for using the information), reasoning (using rules to reach approximate or definite conclusions) and self-correction. While AI is around for decennials, recently it has gained attention and is commercially exploited for all kind of "intelligent" services. It has become a buzzword that obscures reality.

Intelligence has to do with data acquisition and the ability to interpret it; critical reasoning is not required. How test cases shall be generated without reasoning seems rather incomprehensible.

However, AI in testing can do what AI always does: collect and exploit data, classify it and interpret it in view of known pattern. AI does not replace skilled testers, it is not capable of finding new insights or cool new ways of validating software, but AI can industriously generate and compare test cases where people fail because of the hardship. We generate test cases with help of combinatory algebra.

5-2.1 Testing Blockchains

We have seen in *Chapter 1: Why Autonomous Real-time Testing?* how test cases can be represented by the combinatory algebra of arrow terms. We left there with the general statement that arrow terms (1-1) represent test cases, provided the base language $\mathcal{L}$ consists of assertive statements about test cases. The basic arrow terms have an arbitrary but finite number of test data left and a test response on the right. Arrow terms can be combined (1-4), quite similar as test cases also can be combined.

For test automation, it is best to use arrow terms as a combination of elementary arrow terms, each representing one data movement only, and combine them for each object of interest that is touched by some specific test case. Thus, instead of applying equation (1-4), we use a sequence of arrow terms that, when combined, together yield the test case within a test story. Such a sequence of arrow terms is called *Testing Chain*. If all objects of interests touched by the test case are considered separately in an arrow chain, none ignored, the chain is called *Testing Blockchain*. Because, as a matter of fact, this represents a blockchain (Wikipedia, 2018); only, it links data groups – blocks – within a test case instead of encryption keys. A testing blockchain is sort of white-box test: if you execute a test case and trace it with the objects as debugging points, you get the testing blockchain.

Remember the definition of arrow term application in section 1-2.4: *Arrow Term Notation*. The testing blockchain is – after application – the left-hand side of the original arrow term. Thus, a testing blockchain contain more information than the resulting arrow term, or test case, after concatenation using equation (1-6). It contains the trace of the test case.

5-2.2 Measuring Test Size

Arrow terms represent test cases, and test cases can be combined. It is straightforward to represent testing blockchains as a sequence of arrow terms of level one; each block contains one data movement within a test case. The total test size of the test case therefore is equal to the number of blocks within its testing blockchain.

The size of an arrow term is defined such that it still reflects test size. For this, it is not useful to count recursive elements, but only those elements that relate to the base predicate elements, and thus represent executable, testable terms.

As before, let $\mathcal{L}$ be the base language consisting of assertive statements about test cases.

We define the *Size of an Arrow Term* as shown in equation (5-4):

$$|a| = |\{\delta_1, \dots, \delta_n\} \to \rho|, \text{if } a = \{\delta_1, \dots, \delta_n\} \to \rho, \text{where } \rho, \delta_i \in \mathcal{L}$$

$$|b_i \to a| = \sum_i |b_i| + |a|, \text{for all } a, b_i \in \mathcal{G}(\mathcal{L})$$

(5-4)

$$\left|(b_i \to a)_j\right| = \sum_j |b_i \to a|_j$$

The size of a test does not increase with the level. It describes the executable size.

Data movements that appear in more than one test case are multi-counted. Test size depends from the number of test stories, the number of test cases per story, and thus is much larger than the software's functional size. The ratio between test size and functional size is the *Test Intensity*; see section 2-5: *Test Metrics for the Navigator Application*.

Increasing test size is the best way of finding additional defects; however, it does not guarantee it. If there is no customer need, or compliant user story, that let us classify a feature of the software as unwanted, as a "defect", no test will eventually recognize it. *Defect Density* in turn is not affected by increasing test size, because defect density is the number of defects divided by functional size, not test size.

5-2.3   Data Movement Maps and Testing Blockchains

Let $a_1, a_2, \dots, a_n$ be a testing blockchain and let $a_1 \bullet a_2 \bullet \dots \bullet a_n = a$ be its combination. Then, a is an arrow term that represents a test case of size n. This is an immediate consequence of equation (5-4). Testing blockchains thus are something like the "natural" representation of the test case, also encoding the objects of interest that the test case needs to execute in the data movement map.

For instance, assume a test case for the ADAS example has the form (Figure 5-1):

Figure 5-1: Test Case for Testing Route Alert

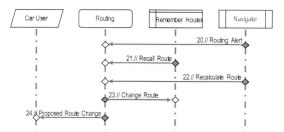

Test cases can be split and joined in the data movement map. Nevertheless, a test case is closely linked to its data movements and therefore also to the data groups and objects of interest. During execution, test cases can be traced in a data movement map. Since ISO/IEC 19651 defines the same measurement rules for functionality as for test, we have the necessary metrics framework for test automation.

The relevant test assumptions for the whole test case are the three arrow terms (5-5):

$$\{Traffic\ Jam\} \rightarrow Proposed\ Route$$

$$\{Icy\ Road\} \rightarrow Proposed\ Route \qquad (5\text{-}5)$$

$$\{\ Heavy\ Rain\} \rightarrow Proposed\ Route$$

Then the corresponding testing blockchain consists of all test assumptions that can be made for the data movements needed to execute these tests (Figure 5-2):

Figure 5-2: Testing Blockchain for the Alert Test Case

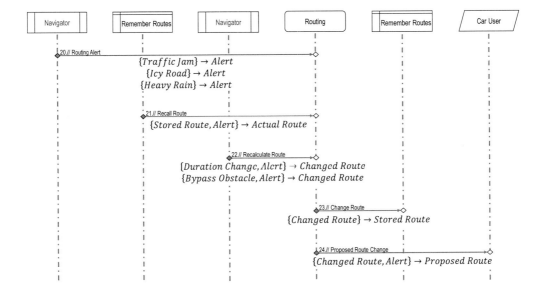

It can easily be verified that the concatenation of elements from these five groups of blockchain arrow terms yields the original three test cases. However, you can combine them in different ways and thus get either traffic jam alerts causing a changed route, or weather conditions doing the same. Response might become different depending upon the cause of the alert.

5-2.4 Using Data Movements Maps to Generate New Test Cases

Testing blockchains have a unique relationship to some path through the data movement map. This makes it possible to use the data movement map for searching variations of existing test cases. Variations can be made as follows:

A-1 Replacing existing test data by a variation of that test data; thus, exploring limits for the controls.

A-2 Tracing back data movements that contribute to some of the data groups addressed in the arrow term; thus, replacing fixed test data by calculated data; adding additional controls, or replacing existing controls.

In both cases it is unclear whether the response of the test case changes as well; executing the test case possibly yields another response. The testing system must learn whether this response is acceptable or not.

For learning, the system has various choices:

B-1 One is by simulating the physical impact the response has. If the response is speed, acceleration or braking, the simulation can predict the possible impact against obstacles.

B-2 Another is using a risk function. If the risk increases above a threshold level, the response is inacceptable.

B-3 Yet another is asking a human tester. Since generation and evaluation of new test cases happens under supervision, not autonomous, humans can decide about the response.

Learning requires that a testing system has a model of the domain under test that allows to judge about the suitability of a response. Such models are sometimes available – e.g., for car driving, accelerating and braking – but sometimes they require human expertise. A car driving in mixed traffic depends not only from its own controlled actions, but also from the perception other road users get. Today, pedestrians look car drivers in the eye to see if they have been noticed. With autonomous cars, this is impractical; car users sitting in the car and playing games or texting have no immediate impact at what the car does next. The best way of learning is to train a neural network for situations, where the car should lower speed at an early stage to make it clear that it grants the right of way, against other situations where denying it is safer. A horn signal would then be more appropriate as a response.

However, with training a neural network, we run into another problem of testing: the neural network changes its behavior while learning. If it learns "on the road", it can unlearn as well. Without continuous testing, the autonomous car, or the ADAS, might unexpectedly fail on challenges that it used to master, initially.

Combining the variations A-1 and A-2 with the variety of responses as outlined in B-1 to B-3, yields the following framework for automatic new test case generation, see Table 5-3. The controls are the test data; the response is the test result. However, we need a mechanism to limit and guide growth of the test suite.

Table 5-3: Automatically Generation of Additional Test Cases

C-1	Level 1: Parametrization of same controls $x_1, x_2, \ldots, x_n$; same response y	Existing test cases without changing logic, changing test data only
C-2	Level 2: New controls $x_1, x_2, \ldots, x_n, x_{n+1}$, same response y	New controls with new test data but response as before
C-3	Level 3: Same controls $x_1, x_2, \ldots, x_n$, new response y'	Same controls with new test data generate new response
C-4	Level 4: New controls $x_1, x_2, \ldots, x_n, x_{n+1}$, new response y'	Same controls with new test data generate new response

5-2.5   MONITORING THE TEST COVERAGE MATRIX

The mechanism to limit and guide growth of the test suite is monitoring the convergence gap on the test coverage matrix. Whatever new test case is selected, it is entered in the test coverage matrix and affects the convergence gap. This is the laborious part of the learning: adding a test case alone almost certainly open the gap, while adding two or more test cases at different cells might well improve the gap.

Thus, there is nothing than try and error, except the case that a sensitivity analysis for the test matrix can be conducted. Such an analysis would allow predicting where to look for additional test cases. Otherwise, we must try new test cases as needed to improve the convergence gap. Since the content of all cells are data movement counts, the more cells a matrix has, the more finely the convergence gap can be adjusted.

However, this is only true if the test stories and the user stories are not linearly dependent. Such situations are detectable with linear algebra. Since all matrix cells contain positive integers only, the matrices usually meet the conditions for the *Perron-Frobenius* theorem; thus, the principal eigenvector exists.

5-3 THE TEST CASE GENERATOR

Remember that test cases are arrow terms containing testing blockchains in their left-hand side. Thus, the *Test Case Generator* has access to the full testing blockchain, and it needs that information. The data movement map in Figure 5-4 designs a test case generator. We comment on the six functional processes.

Figure 5-4: The Test Case Generator as a Data Movement Map

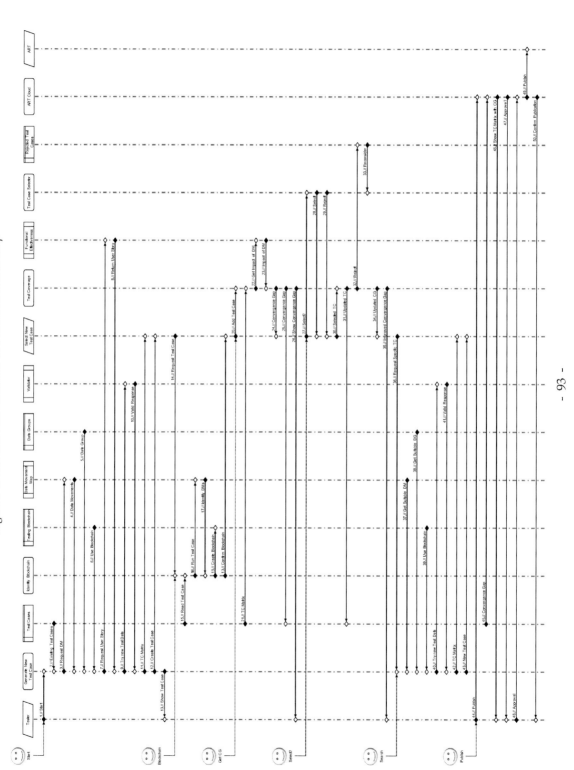

5-3.1 Start Generating a New Test Case

The first step in Figure 5-4 is by using the testing blockchain to create a new test case.

First, get the existing test cases for all test stories involved. Next, collect the data movements executed by that test case. This yields the candidate data movements for the test coverage matrix.

The data groups are needed to build the testing blockchain. By recombining the arrow terms inside the testing blockchain, several new test cases can be generated. However, their response is not given; it must be asserted by some validation application that might involve human judgement. As a result, the test case has now a valid response; otherwise it is rejected. The functional process ends with announcing new test cases to the device that selects those test cases which have the potential to lower the convergence gap in the respective test coverage matrix.

This functional process uses information from *Functional Effectiveness* as heuristics which test cases to generate. This allows to generate test cases that support certain user stories; for instance, those that lack support in the test coverage matrix. Such a functionality speeds up the test case generator but also can block finding useful other test cases that are not obvious. A random generator must ensure the necessary fuzziness.

5-3.2 Calculate the Convergence Gap

Calculate the convergence gap for an updated test suite, creating the test coverage matrix and using equation (2-1).

Figure 5-5: Calculate Convergence Gap

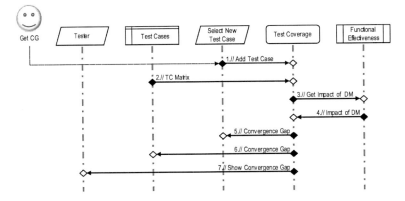

Calculating the convergence gap is straightforward. As before, calculation considers functional effectiveness for counting the impact of each cell in the test coverage matrix.

A test coverage matrix is represented by selecting relevant test cases within the test stories. Many candidate TC matrices will be needed for the selection step (5-3.3), coming next.

5-3.3   SELECT THE NEW TEST CASE FOR INCLUSION INTO TEST COVERAGE

Select the new test case for inclusion in the test coverage matrix, based on the convergence gap of the test coverage matrix (Figure 5-7). This step selects or rejects test cases for inclusion into the test suite, and consequently the test coverage matrix. The decision depends from the convergence gap, computed before. The tester remains informed.

Test Case is abbreviated by "TC".

Figure 5-6: Select a New Test Case for Inclusion into Test Coverage Matrix

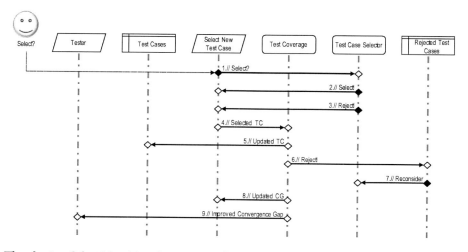

The device *Select New Test Case* is an information exchange bus, triggering the necessary steps to create and select a new test case, and decide whether to include it into the test coverage matrix and thus the test suite.

5-3.4   SEARCH FOR A NEW TEST CASE

The test coverage matrix might still not be satisfactory. Sometimes, it is necessary to search for a new test case that adds impact to specific test stories or user stories, based on specific data movements that add weight to some weakly supported user story. This functional process is called when needed.

The following Figure 5-7 demonstrates how such test case creation can be initiated and achieved. The functional process *Generate New Test Case* uses the object *Data Movement Map* and the *Testing Blockchain* to try a new test case.

Figure 5-7: Create New Test Case Executing Specific Data Movement

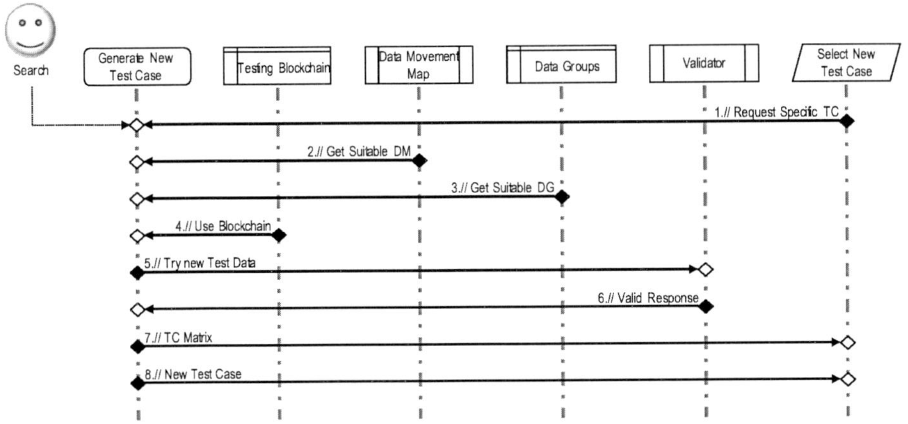

As before, the *Validator* application is needed to validate the response in the test case. The device *Select New Test Case* remains in control for the enhancement of the test coverage matrix, and thus the total amount of test cases per test story.

5-3.5   CONCATENATING TEST CASES

Often test coverage matrices have a strong diagonal bias. This is caused by test stories that reflect only a few user stories and neglect the interactions between "far-away and unrelated" user stories.

Unfortunately, it is exactly with such cases that serious defects pop up that otherwise remain undetected. An example from airline industry are sensors, that seem to work fine as specified but relate wrong measurement data. This is hard to get covered in manual tests, because the supplier of the sensors often is different from the organization that programs flight controls. In fact, airplanes, autonomous cars, and many more software-intense systems consist of thousands of independent pieces of software, for instance distributed among *Electronic Control Units* (ECU), but somehow interacting with each other (Ebert & Jones, 2009). See section 6-3 *AHP for Testing* for a sample case.

The result of such situations are test coverage matrices with a lot of empty space, not testing how user stories for the one kind (e.g., sensors) impact other software (e.g., flight control). The method of choice for spare test coverage matrices is to concatenate test cases using equation (1-7) respectively (1-8). This concatenation preserves information about the individual tests being concatenated, and thus remain executable.

5-3.6 PUBLISH TEST SUITE

If the tester is satisfied with the result, she or he publishes the new test suite to the cloud for use by all connected ART clients (Figure 5-8).

ART users can now download the test suite, execute the tests and upload the test results. These results might be consulted in case of failure or incident, to assess responsibilities of software suppliers, or at least to learn how to make the software better.

Control is given back to the (human) *Tester*.

Figure 5-8: Publish Test Suite to the Cloud

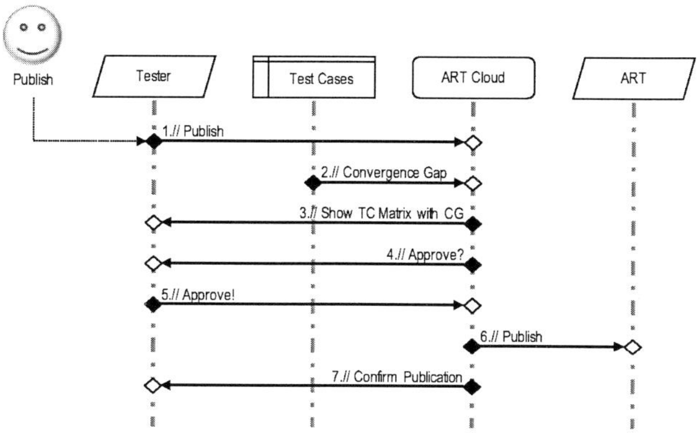

5-4 THREE STANDARD TESTS

In preparing *Autonomous Real-time Testing* (ART), three standard tests are used to protect a software-intense product against privacy violations and the consequential safety risks:

- The *Data Walker Test* (DWT) consists of visiting all objects, listing their published methods and assessing their privacy protection status. Data groups, retrieved from the model, are used to detect hidden interfaces, by checking whether those data groups appear in other objects. If they are not, there must be data movements that are not listed in the model.

- Each data movement is assessed in view of its privacy protection needs whether it is effectively protected. This yields the privacy protection index and is called the *Data Movement Test* (DMT).

- The *Sniffer Dog Test* (SDT) is one layer below the application and monitors data communication traffic. Each data package must be assignable to some data group of the model.

5-4.1 THE DATA WALKER TEST

The DWT is basically a static test, if source code is available. Interface specifications are good enough. If not available, the DWT walks the data movement map model, trying to visit each identified object of interest. If the software supplier provided for such testability, a list of public methods is offered that can be used to execute the visit. If not, effective DWT testing depends on the ability of the tester to model the functionality with a suitable data movement map, plus how well those objects of interest effectively can be visited. For the visit, they need to exhibit some programmable interface. Else the DWT is difficult and eventually the privacy index cannot be determined.

However, in those times of open source computing, the DWT test can quite often be executed, and it should be a normal requirement for an Original Equipment Manufacturer (OEM) that he needs to provide equipment that is DWT-testable.

The test runs as follows:

1) Identify all *Data Movements* that go out or into the object.
2) Determine the *Data Groups*.
3) Compare with the *Privacy Needs* for these data groups.
4) Compare with the *Safety Impact* for these data groups.

The privacy needs and the safety impact are attributes to the COSMIC model as explained already in section 4-2. In short, the left-hand part of the privacy assessment and the right-hand side of the safety risk assessment. The test detects typical failures such as data movements moving data groups without privacy needs attribute, or without safety impact attributes.

To some extent, the DWT is a model validation test. However, analyzing the code or the behavior of the object of interest does also detect data movements that are not part of the model, for technical reasons for instance. Some data movements remain invisible to the functional user. They are not required by a FUR. Thus, not all findings of the DWT are automatically data leaks, but they should be investigated whether they have such potential.

Obviously, it is also possible that the model is not capable and comprehensive according ISO/IEC 14143, or not all functional users have been taken in due consideration. In both cases, the results of the DWT might cause rework and fixes, be it to code, to the embedding container, or to the model itself.

5-4.2 THE DATA MOVEMENT TEST

The DMT is the logical continuation of the DWT: all data movements found by the DWT are tested against effective protection. This is a dynamic test. Usually it is expected that data is encrypted according some one-way or two-way protection scheme. Although this could be a static test, if code is available, normally such a test must be executed dynamically, looking at the data moved whether it is readable without encryption key or not, and where the key originates.

The most efficient way to execute a DMT is by running the software in some standard environment and tracing each data movement executed. The frequency of execution is also measurable; thus, it serves as well for assessing *Incurrence Frequency*. From its results, both the *Privacy Index* and the *Safety Risk Index* can be calculated.

The DMT does not validate the model but its implementation.

5-4.3   THE SNIFFER DOG TEST

The SDT is a black box test looking at the dynamic execution of the software. It monitors all communication channels that are used by the software. It expects each data communication matching one or more data groups identified in the model. If some data communication does not fit into the model, it might indicate an illegitimate data movement, or a shortcoming of the model.

The SDT needs access to keys used for encryption and therefore can be executed in combination with the DWT, and thus complements model validation.

5-5 THE DEVOPS PARADIGM AND SOFTWARE TESTING

The DevOps paradigm requires that software development interacts with operations, and it is not called *DevTstOps*. Testing is part of product development or not done at all. That the tendency is for "not done at all" is more than obvious. Untested software publishes today's newspaper, runs train systems, and delivers organizational schedules, making every aspect of our life more and more adventurous.

Thus, modern software testing must become part of the operation of software, not only part of software development. This means, software must be able to test itself at any time and occasion. Automated tests must be built into the software, and available for execution to both consumer and supplier.

Agile software development had developed a branch called Test-Driven Development (TDD) that creates unit tests before delivering any functionality. Unfortunately, and unnecessarily, these unit tests usually become not part of the delivered code, possibly for fear of decreasing performance. But performance is not a major issue nowadays and is only affected when the software starts testing itself while it should be available for performing its primary purpose. Obviously, a software-based system can cope with such a constraint.

Test stories and test cases can be stored in any software and can be executed at any time that the workload permits. Hence testing must be fully automated. This is still difficult but state-of-the-art. And if performance still matters, missing computing power can be borrowed from cloud systems.

5-6 THREE INNOVATIONS NEEDED

The current art of testing is outdated. As already stated, the ISO/IEC/IEEE 29119 testing standard (ISO/IEC/IEEE 29119-4, 2015), part 4, identifies 23 different so-called *Test Coverage Items*, but not software functionality. As if software functionality were not items in software that can be well distinguished and handled.

While non-functional software characteristics exist that can be tested, dynamic test is per se functional; otherwise it would be static testing. While static testing, e.g., code analysis, is highly important for technical debt and for safety and security assessments, static testing never suffices to ensure proper functioning of mission-critical software.

But dynamic testing of complex systems inclusive artificial intelligence requires three innovations.

5-6.1 FIRST INNOVATION – TEST AUTOMATION

The first innovation needed refers to *Test Automation*. Traditionally, tests were successful when they produced reproducible responses. Reproducible responses cannot be the goal of testing in learning systems. We therefore propose a new method of specifying test cases using *Combinatory Logic*. This is a system that maps preconditions to postconditions expressed by formulas. It classifies similar test cases. A test is passed when the response formula is found to be true. Determining whether a response is valid or not might be delayed until running the test.

For autonomous cars, such test conditions and test responses fit well. Things like speed limit, speed range, acceleration and breaking effectiveness can be better expressed with formulas, referring to some thresholds, rather than by fixed test data, referring to known, expected and correct responses. Varying road conditions or truck load loads can influence the correct answer in a way that is hard to predict.

For test automation, we refer to *Data Movement Maps* that describe a software in terms of data groups being moved from one object of interest to another. These objects, be it functional process, device, other application, or persistent store, all need being equipped with *Test Stubs*. Test stubs are the pieces of code that emulate a device, or other application, in a physical environment. Persistent stores and functional processes also need test stubs; in cases where some fixed behavior is expected in the test case. In case of hardware in the loop, we effectively call for a *Digital Twin* (El Saddik, 2018).

Simply speaking, test automation means programming test stubs such that they execute certain test cases. This is what makes ART possible, at the end.

5-6.2 SECOND INNOVATION – TEST METRICS

The second innovation needed refers to *Test Metrics*. Test metrics must be independent from implementation, especially from code, as code for certain services needed by the system under test are often not available, and code size is irrelevant. Test metrics like test size, test intensity, test coverage and defect density must compare with functional size. It is the functionality that's being tested, not code. Moreover, test metrics must be understandable by consumers using a software-intense system, like ecolabels for today's products.

Consequently, test metrics must refer to functionality in use, and not to obsolete requirements or specifications. Test metrics cannot refer to code, as code is usually not available for measurement, be it that functionality originates from cloud services or proprietary code.

Moreover, certain code today is self-correcting and usually not responsible for functional failures. Code is not the object of testing; it is the systems functionality.

Test metrics must use the same measurement method as functional size metrics. We can use the same data movement maps for representing pieces of functionality as we use for tests.

5-6.3 THIRD INNOVATION – ART

More challenging is adapting test stories and test cases continuously, by new experiences made by the software, changing the behavior of the complex system. The software might modify itself, or modify data that controls its behavior, or the system might encounter new situations in changed environments. For instance, an autonomous car that encounters new traffic situations and learns from them might cause the controlling software to behave differently than before. Test cases, and even test stories, must adapt. An automated test repository is needed that grows with the changes to the software, and with additions to the system. This is the essence of *Autonomous Real-time Testing* (ART).

This is the major innovation that we propose to software testing. To make it work requires even more innovations. Future software contains its own testbed that users can run anytime when needed and see the result. Moreover, the software can run the tests autonomously, for instance when encountering new situations with an autonomous car, or when adding or removing system components such as an IoT device, or when adding a new truck member to a truck platoon, or when commissioning a new software-intense train system. Even when establishing communication with another car or road user, a short test might be appropriate to establish trust into the new relationship and the communication means. ART also regularly checks existing software for newly introduced software faults, vulnerabilities, changed features, or hardware wear such as breaking effectiveness.

CHAPTER 6: TESTING HIGHLY COMPLEX TECHNICAL SYSTEMS

The problem with complex technical systems is testing. Testing is utterly complex and sometimes not feasible because of the many subsystems involved. People cannot devise enough test cases because they cannot test everything against anything.

Moreover, if your functional size exceeds, say, a million – this is easy for airplanes or spacecraft or even for autonomous vehicles or trains – you need a test size of ten to hundred million for achieving a reasonable test intensity. Humans cannot deliver that. We need machines to do this.

6-1 TESTING DIGITAL TWINS

Whenever testing software-intense systems, testing with *Hardware-in-the-loop* can be quite demanding. The hardware needs to put in a state that produces the required test data. In many cases, this is impossible or very costly to keep hardware in the loop while testing large, complex systems.

As already explained, we rather test *Digital Twins*, where hardware components, sensors, and actuators are emulated rather than tested in the loop. Digital twins today are available for all kind of hardware component build into software-intense systems.

6-1.1 THE DOUBLE-TIDDLEMUTZZ EXAMPLE

The *Double-Decker Tilting Long-Distance Multiple Unit Trainset* (D²TLDMUTS) serves as an example to explain the new concepts. D²TLDMUTS is pronounced "*Double-Tiddle-mutzz*", with a sharp "zz" hiss at the end. It refers to a large Intercity multiple unit trainset, able to run on international railway traffic as a double-decker with restaurant, with children's corner, offering space for people with disabilities, featuring roll compensation for faster driving around a curve, comfortable enough for three to six hours of daytime train riding. Thus, responds to many different – and conflicting – needs.

It has been ordered by a European railway operator, originally targeted for 2013 but now, in autumn 2019, finally being commissioned. Commissioning started in February 2018 and will last well into 2020. Normally, commissioning a train takes three to six months; assuming, it is a commuter train with mostly standard components. But this train is utterly complex. After the first year of commissioning, the number of bugs found, and problems encountered, piled higher than ever. Suppliers and train service operator realized that they are only half-way through before letting the D²TLDMUTS run operational services. Such kind of failure is common not only with train operators; several similar cases occurred in the last few years in aircraft industry as well and is likely to happen with autonomous cars; let alone pharmaceutical industry.

The problems encountered with the D²TLDMUTS are basic: it is virtually impossible for humans to create complete test suites for such a complex, software-intense system. Consequently, commissioning such a train set takes very long, much longer than ever planned. Defects touching across the various systems are detected in this trial period only. This is very late, because every modification of train software requires an extra re-certification and a new admission procedure.

6-1.2   COMMISSIONING REPLACES TESTING

Key of testing complex systems is understanding the needs (or values) of the train operator, in our case, or the needs of the customer, in general. The needs of the train operator are the key means for distinguishing relevant test cases from unnecessary tests, allowing test case automation and finally *Autonomous Real-time Testing* (ART).

Commissioning such a software-intense system takes an unpredictable amount of time. Not only due to the difficulties of designing such a multi-purposed system – even if the supplier did an excellent engineering job – but far more in the commissioning of its software. Either instrumentation and control fail, or the door control stops working, or you cannot connect to the *European Train Control System* ETCS (Wikipedia, 2019). If the software somewhere fails, the only remedy is to switch everything off and then reboot the train. This takes ten to twenty minutes. In rail networks like in Switzerland, the Netherlands or Japan, after such a reboot, the timetable is out of control for the rest of the day; nationwide. A software breakdown during train operation must never happen; this is an absolute constraint.

It is unknown how big the software is; probably, even the supplier does not know. This is scandalous. Today, publishing software size seems nothing aimed at the public, and train manufacturers still do not behave as a software house, although they are.

However, if we assume 500'000 CFP, we might still underestimate the complexity of a D²TLDMUTS, with instrumentation and control, with information and ticketing services for the public riders, incorporating services needed to control and minimize energy consumption, comfort services controlling all the technical installations on board, including heat control and air circulation, and all the recording needed for the big amount of data. It is not a simple commuter train, or a locomotive hauling trailers, the D²TLDMUTS is a multiple unit railcar with restaurant, children playing area and space accommodating a thousand passengers, including people with disabilities.

It is impossible to let testers set up enough test cases, manually. Too many systems interact. It is a typical case for *Combinatory Logic*. Test cases must be created automatically, combined from test stories with basic test cases. Such tests can run, searching for weaknesses and bugs, before commissioning the train, or put the system into service.

Testing such a system involves several steps. Note that we do not need textual specifications. Although in theory specifications would be helpful to set up test stories and the related test cases; in practice, specifications are meddling up the important with the marginal and thus of limited value. In any case, specifications without priority profiles are near to useless. No written document can describe adequately the complexity of our D²TLDMUTS train system in full.

6-2 THE FUNDAMENTALS OF TESTING COMPLEX SYSTEMS

Traditionally, the customer needs are what matters and defines the goals of testing. However, when buying train sets, the customers are not primarily the train riders but the train operators. It is the train operators' interest that the trains run on schedule and its riders come back again, remaining loyal customers of the train operator. While train riders and operators might share common values, in some other respects they differ. Train riders do not care much about the costs of running the trains reliably; in turn, operators do. Operational cost must remain below older trains.

Figure 6-1 shows *The Complete Analytic Hierarchy Process for the D2TLDMUTS*. The lists the *Operator's Needs* regarding the new D²TLDMUTS is hierarchically grouped and analyzed using the *Analytic Hierarchy Process* (AHP). This time AHP in full, with one level of hierarchy. The hierarchy reflects those subsystems of the D²TLDMUTS that we intend to test. For testing, each group will need its *Operator's Needs* for defining the goals of tests. All group tests combine for the full D²TLDMUTS tests, letting ART filling the test gaps in between groups.

Figure 6-1: The Complete Analytic Hierarchy Process for the D²TLDMUTS

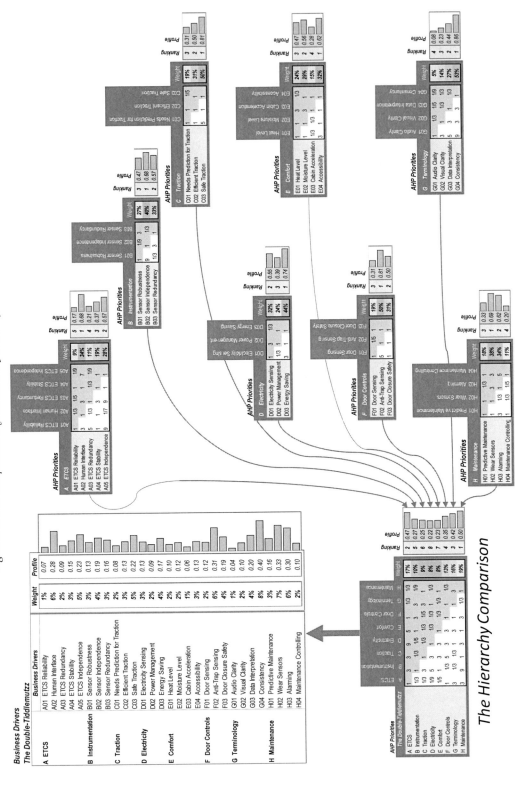

With a few more textual attributes to explain what is intended by the Operators' Needs for the D²TLDMUTS (Figure 6-2):

Figure 6-2: Operators' Needs for the D²TLDMUTS

Business Drivers			Attributes	
A ETCS	A01	ETCS Reliability	Safety for humans	Safety for instrumentation
	A02	Human Interface	Communicate clearly	With operators and passengers
	A03	ETCS Redundancy	All ETCS equipment is redundant	If results differ, alert!
	A04	ETCS Stability	Unambiguous status	Consistent
	A05	ETCS Independence	Each subsystem is autonomous	Can close or fail
B Instrumentation	B01	Sensor Robustness	Legibility	Completeness
	B02	Sensor Independence	Save energy	Use energy wisely
	B03	Sensor Redundancy	Have two sensors where applicable	Compare sensor data
C Traction	C01	Needs Prediction for Traction	Predict knowing train load	Predict knowing weather conditions
	C02	Efficient Traction	Optimize acceleration	Minimize energy consumption
	C03	Safe Traction	Safety for humans	Safety for instrumentation
D Electricity	D01	Electricity Sensing	Sensing the power supply	Adaption traction
	D02	Power Management	Distribution of power in train	Laptop plug supply
	D03	Energy Saving	Extract relevant data	Keep data for analysis
E Comfort	E01	Heat Level	Convenient for passengers	Both women and men
	E02	Moisture Level	Convenient for passengers	Enough dry
	E03	Cabin Acceleration	Convenient for passengers	
	E04	Accessibility	Entrances	Toilets
F Door Controls	F01	Door Sensing	Door knows who's inside	
	F02	Anti-Trap Sensing	Doors must not close by force	Avoid dangerous conditions
	F03	Door Closure Safety	Each subsystem is autonomous	Can close or fail
G Terminology	G01	Audio Clarity	Understandable	Also for the hearing impaired
	G02	Visual Clarity	Legibility	Completeness
	G03	Data Interpretation	Doors must reopen when needed	People must never get trapped
	G04	Consistency	Consistent Messages	Adaptive terminology
H Maintenance	H01	Predictive Maintenance	Alert well in time	Before failure
	H02	Wear Sensors	Put sensors near wearing equipment	Have sensors for all wear & tear
	H03	Alarming	Timely alarms	Alert in case of uncertainty
	H04	Maintenance Controlling	Make sure maintenance is effective	Also check efficiency

6-2.1 THE HIERARCHY OF OPERATOR'S NEEDS FOR THE D²TLDMUTS

Each new, complex, system requires training, adaptation of operational processes and new standard procedures for operations and maintenance. For instance, older electric traction gear only needed a switch being turned off for putting them out of service, while modern equipment has many functional processes that need being shut down in an orderly manner. A train software feeding a data base might cause database corruption when turned off unexpectedly; after this, restarting software plus database might take a long time because the database needs being repaired. Locomotive engineers might not be used to such thinking; thus, they need training and instruction to understand new technologies. On the other hand, software engineers that program instrumentation and control are probably not aware of the operational conditions and constraints. Thus, they take things for granted that are not. The standard approach to such a problem is *Quality Function Deployment* (QFD); see (ISO 16355-1:2015, 2015).

The method of choice to find priorities is the *Analytic Hierarchy Process* (AHP). It makes sense to do the pairwise comparison once per needs' group and combine their profiles. The result is quite surprising. While *H02: Wear* Sensors, *F02: Anti-Trap Sensing* for doors, and *H03: Alarming* clearly dominate other needs; the need for unambiguous communication *G04: Consistency* wins over all. This is a clear indication where the software problems arise: lack of consistent communication between the many electronic and software components in the train sets.

For a train set that assembles components of various suppliers with software developed during different ages, consistent communication is not something for free, but something that requires decent consideration and dedicated work. The components of the D²TLDMUTS originate from different ages and suppliers; regulations have changed over time and with regulation terminology, the meaning of terms.

6-2.2   TERMINOLOGY MANAGEMENT

These requirements are relatively new. However, since a few years the discipline of *Terminology Management* has evolved responding to the needs of the European Union. This suggests developing a *Terminology Broker* that not only controls, but also consolidates and levels out the different messages obtained from instrumentation and controls with those from the signaling system and from traction. Such a terminology broker also enables testing and has a few more advantages (Cabré Castellví, et al., 2017). Setting up a learning system that learns how to interpret the thousands of messages coming in from the various components is probably the simplest way to create a terminology broker for such a complex software intense system.

For many readers, it might not be clear what a terminology broker is. Basically, it is a message broker that "understands" messages and can translate a term from one environment into the correct term in a different environment, translating the meaning unambiguously. *Terminology Management* is a relatively new language science (Fathi, 2017) aiming at providing a platform for technical and societal communication among members of different communities such as within the European Union. Terminologists establish the terms specific to a field of activity, define them, and then find equivalents in another language. They also define the terms in use for businesses, databases, glossaries, dictionaries and lexicons for the purposes of standardization.

6-2.3   THE ANALYTIC HIERARCHY PROCESS

The effect of this AHP (Figure 6-1) is stunning; it is an eye-opener. While everybody probably would agree, without hesitation, to the principle that AI could help with complex technical systems, the idea that AI could provide a terminology broker functionality is a somewhat surprising consequence from the 29 different operators' needs. While these needs look complicated enough to handle, this sample size still is quite below reality and we do not try to make it more detailed; otherwise, it would not fit into this book's format.

On the other hand, while quality or marketing managers are tempted to concentrate on the 7 ± 2 most relevant needs (Gigerenzer, 2007), technical people must concede that needs not carried forward into programming probably will also not be tested. Thus, complex software-intense systems clearly require other teaching methods than examples in a traditional book.

For people not familiar with the *Analytic Hierarchy Process* (AHP) we give a short explanation how to read Figure 6-1. The basic principle of AHP is pairwise comparison among comparable criteria. Therefore, the evaluator must compare each criterion with each other. However, to reliably compare 29 criteria with each other is difficult if not impossible.

Saaty therefore introduced the AHP. The AHP uses Euclidian vector space metrics – the direction of unit vectors that we call *Profiles* – to compare two evaluations. This allows splitting these comparisons into smaller groups according a hierarchy. Because the result of comparisons are profiles rather than linear weights, you can combine such profiles simply by multiplication. Profiles, as already explained in section 2-3, define a direction within an event, or in this case a decision room and combining directions is possible without introducing a bias for some of them. The *Hierarchy Comparison* AHP matrix defines by its solution profile how to combine the priorities of the individual part pairwise comparisons for the full AHP. The components of this profile are used as weights when combining the various part solution profiles from the part pairwise comparisons.

More on AHP can be found from its inventor (Saaty, 1990), or in the precedent book of the author (Fehlmann, 2016, p. 33ff).

6-2.4   THE SOFTWARE UNDER TEST

It is not possible to include data movement maps for the full D²TLDMUTS in this book. However, we have construed the AHP hierarchy in such a way that it maps the part software applications of the D²TLDMUTS. This is obviously always possible, and we can set up user stories and test stories for each of the eight parts; although, we are still oversimplifying. User stories and test stories yield test coverage matrices for each part application. Each part application has its own data movement map, although these applications do talk to each other; thus, have data movements connecting them. The initial test coverage matrix for the D²TLDMUTS is then simply the combination of all nine test coverage matrices, weighted by the profile of the pairwise comparison AHP matrix that governs the hierarchy combination. Multiplying matrices by a linear profile component is a standard operation in linear algebra and yields a linear combination of the other nine test coverage matrices. The combined response profile then matches the profile of the 29 operators' needs, up to some convergence gap.

However, for testing, leaving all the interactions out that occur between the nine software applications would introduce an unbearable safety risk. The gaps can be filed in manually, but better this is addressed by ART. ART does not work on the nine software applications alone but on the whole system: thus, filling up the empty space.

This means for instance that ART adds test cases to test stories that for instance refer to door closure, connecting it to ETCS status. Exactly such dependencies have hit the actual D²TLDMUTS' commissioning. Thanks to ART, such tests can be done before the train operator is involved; and, what is even better, they are generated by a structured, almost "intelligent" algorithm. It does happen according the test generator rules when some data movement exists that connects ETCS information – e.g., free track ahead – with door closure control software.

The details when the D²TLDMUTS can depart or let passengers disembark are modelled in the *Door Control* part application but this application depends in many respects from other part applications – such as *Traction* and *ETCS*. Testing *Door Control* is not complete without taking these interferences into account. But things become complicated with that many interrelated systems; setting up test stories and finding relevant test cases becomes a tedious task.

Below we show two of the part applications – *Door Control* in Figure 6-3 and *Terminology* in Figure 6-4. The *Combination* application has data movements that connect almost all the part applications with each other. This can be used to generate testing blockchains for ART, connecting all different parts of the D²TLDMUTS system to extend test coverage. The *Combination* application is already too large to fit on a book page. Readers interested in these details can study all related data in the shared cloud data accompanying this book (Fehlmann, 2019).

However, the extract shown in Figure 6-5 on page 113 is enough to demonstrate the mechanisms of ART with complex technical systems. By data movements, the *Combination* application connects all other part applications to collect a comprehensive status of the D²TLDMUTS. This part is shown. For constructing the testing blockchains needed to test status, these are the essential data movements.

Figure 5-3: The D²TLDMUTS door control application – opening, closing and locking doors

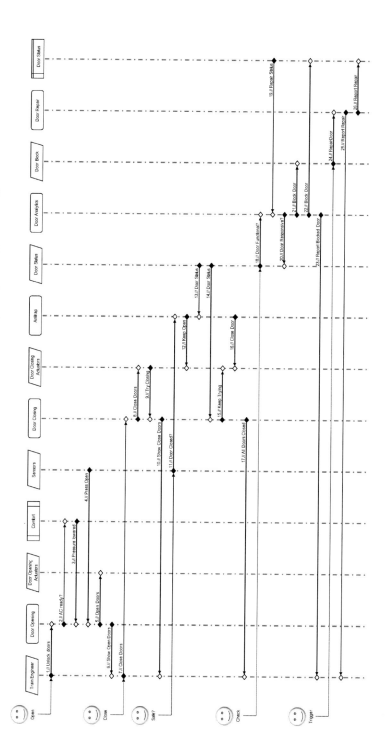

Figure 6-4: The D²TLDMUTS Terminology Application – Test Case Generator as a Data Movement Map

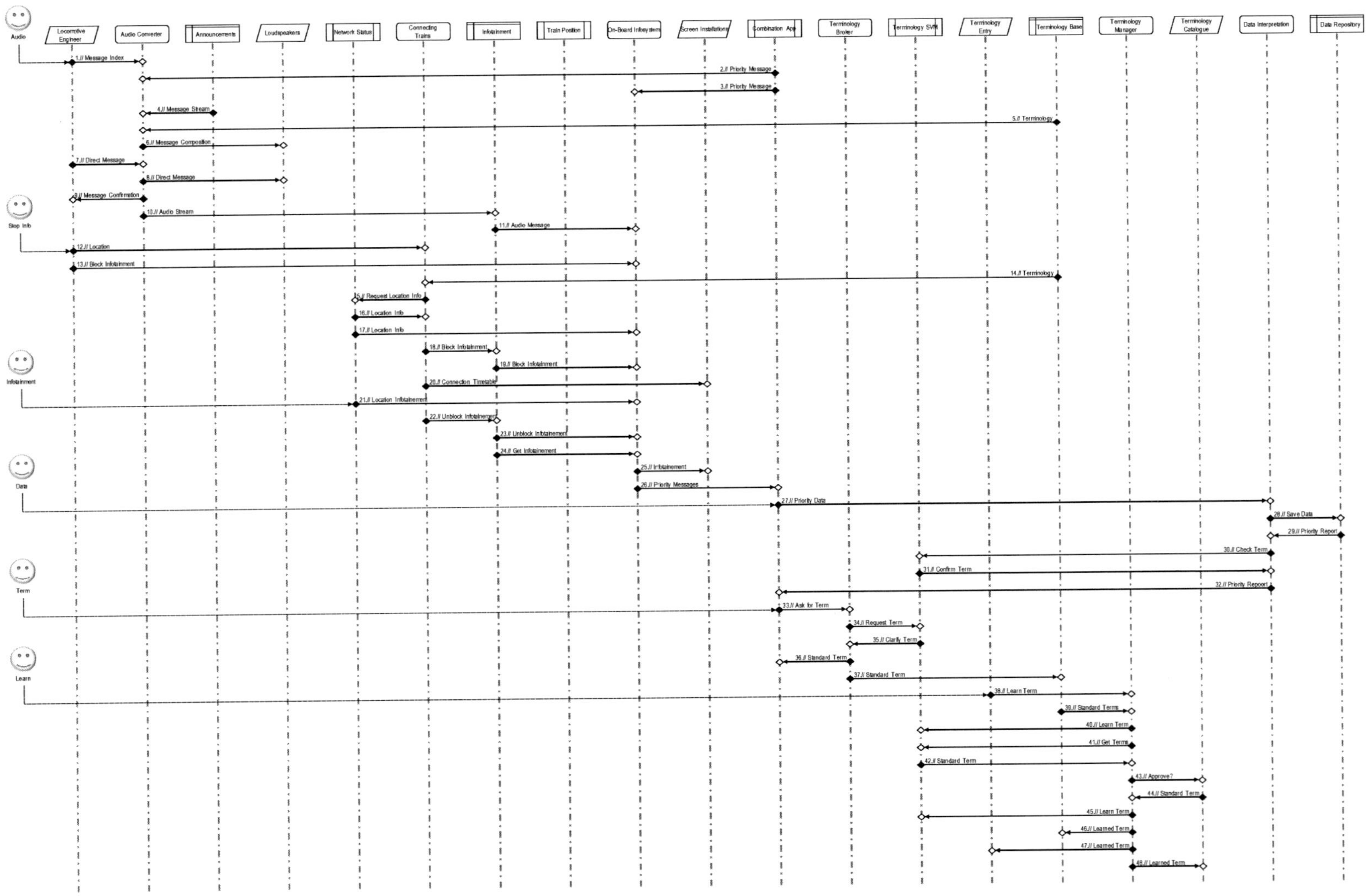

Figure 6-5: Extract from the Combination Application Combining Other Part Applications of the D²TLDMUTS

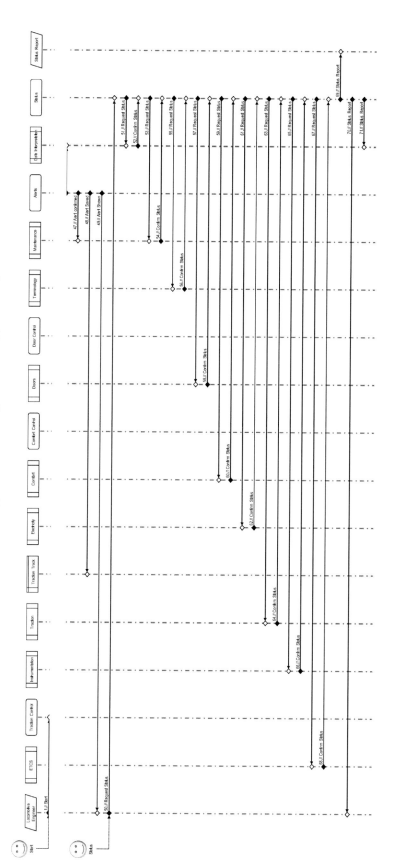

6-3 AHP FOR TESTING

It has already been noted that testing is not very effective without reference to the goals of testing, the needs of the customer, or user. The AHP is the method of choice to analyze and prioritize needs of the customer. However, when used in its full, hierarchical form, AHP is even more helpful. The hierarchy typically addresses system parts; parts can be tested independently, and their test coverage matrices combined the same way as the part AHP. This is quite straightforward but allows managing large test coverage matrices, when combined with ART.

6-3.1 USING THE AHP HIERARCHY FOR SETTING UP TEST STORIES

Let $A_1, A_2, ..., A_k$ be a sequence of AHP pairwise decision matrices with solution profiles $y_1, y_2, ..., y_k$ respectively; $k \in N$; $k > 0$. Thus, up to some numerical imprecision, $A_i y_i \cong y_I$, for $i = 1, ..., k$. because there are no algebraic solutions for Eigenvectors.

Let $\bar{A}$ be the *Hierarchy Comparison* with the solution profile $\bar{y}$. $\bar{\bar{A}}$ *is a* $k \times k$ square matrix; thus $\bar{y} = \langle \bar{y}_1, \bar{y}_2, ..., \bar{y}_k, \rangle$ is the solution profile for $\bar{A}$.

The combined solution profile for the full AHP is shown in equation (6-1):

$$v = \sum_{i=1}^{k} \bar{y}_i y_i, \qquad i = 1, ..., k; k \in N; k > 0$$

$$Full\ AHP\ Solution\ Profile = \frac{v}{\|v\|}$$

(6-1)

Note that equation (6-1) denotes a sum of profile vectors, divided by its Euclidian length; thus, making the result $v/\|v\|$ yet another profile.

According Saaty, this is the mechanism how a hierarchy of decisions should be handled. The key point is using vectors of normalized length that can be added, subtracted, and multiplied by scalars. Intuitively, this represents the direction to take in the decision space, and that is what AHP is all about.

6-3.2 TESTING THE PARTS

Now, each of the A_i pairwise decision matrices describe the needs of the customer with respect to its part, be it ETCS, door control, communications. It is easy to describe the functionality required to fulfil these needs by data movement maps, and verify effectiveness of the implementation by the *Functional Effectiveness* transfer function E_i.

E_i maps the user stories onto customer's needs, by counting data movements needed to implement certain user stories, and thus assigns data movements to user stories. These transfer functions are described by matrices that are not square; typically, many more user stories are needed to implement needs of the customer, than needs itself.

For the user stories, a profile results that describes the importance of the functionality described by the user story to the customer in view of the stated needs y_i. Let u_I describe this profile. Its dimension is the number of user stories needed to implement the topics decided with A_i.

For each of these sets of user stories with profile u_I, an initial set of test stories is needed to cover the user stories with tests, together with an initial sample starting set of test cases. The resulting test coverage matrices F_i map test stories onto user stories, again based on the data movements executed in the respective test cases. In turn, its solution profile we denote by s_i. By definition, $F_i s_i \cong u_i$ holds up to the convergence gap. This nearly equality ensures test coverage for each of the i hierarchical part applications, referring to the A_I pairwise decision matrices for the initial needs of the customer per part application, for $i = 1, \ldots, k$.

6-3.3 DOOR CONTROL

The creative task is inventing test stories and test cases that effectively test the implemented functionality. The initial needs of the train operator help managing the complex system and its setup.

Figure 6-6: Pairwise Comparison for Door Control

AHP Priorities Operator's Needs	F01 Door Sensing	F02 Anti-Trap Sensing	F03 Door Closure Safety	Weight	Ranking	Profile
F01 Door Sensing	1	1/5	1	19%	3	0.31
F02 Anti-Trap Sensing	5	1	1	50%	1	0.81
F03 Door Closure Safety	1	1	1	31%	2	0.50

The following four user stories implement door control (Figure 6-7):

Figure 6-7: User Stories for Door Controls

User Stories Topics	As a ... [functional user]	I want to ... [get something done]	such that ...[quality characteristic]	so that ... [value or benefit]
1) Q001 Stop	Train Operator	open all doors	passengers can leave the train and new passengers can bord	exchange is fast
2) Q002 Start	Train Operator	close all doors	no passengers are trapped in a door	start is fast and according schedule
3) Q003 Safety	Train Operator	get an alert for any door left open or needing repair	I can block defect doors when stopping	passengers recognize door that are out of operation
4) Q004 Pressure	Train Operator	lower the air condition pressure	doors do not produce air blow when opened	cabin pressure does not interfere with door opening

These four user stories explain the basic functionality of door controls and implement the *Door Control* operators' needs effectively, as shown in Figure 6-8:

Figure 6-8: Functional Effectiveness for Door Controls

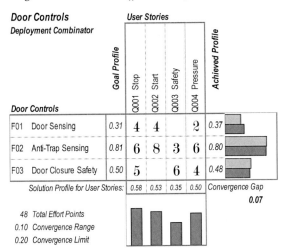

With functional effectiveness, we know which data movement is assigned to which user story and therefore we can calculate test coverage, given a suitable set of test stories, by looking at the data movements executed by the test cases defined per test story:

Figure 6-9: Initial Test Coverage for Door Control

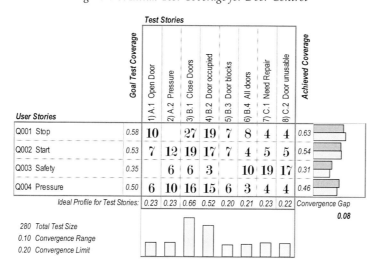

Figure 6-9 explains how to test door controls (the data movement map shown in Figure 6-3). For the details of the initial test cases for the eight test stories addressing door control functionality, we refer again to the shared cloud data accompanying this book (Fehlmann, 2019).

6-3.4 TERMINOLOGY

Note that locking doors has another meaning when looking at door control from the traction or ETCS standpoint than from door control itself. Locking doors – and sometimes even side-specific door locking – is a precondition for the train departing. Unlocking when stopping at stations is necessary for letting passengers disembark.

Thus, terminology plays a role when door control talks to ETCS or traction, and this is why we chose the *Terminology Application* as next part of the D²TLDMUTS software; see Figure 6-4 for the data movement map.

This is the profile of the train operators' needs (Figure 6-10):

Figure 6-10: Pairwise Comparison for Terminology

AHP Priorities	G01 Audio Clarity	G02 Visual Clarity	G03 Data Interpretation	G04 Consistency	Weight	Ranking	Profile
Operator's Needs							
G01 Audio Clarity	1	1/3	1/5	1/9	5%	4	0.08
G02 Visual Clarity	3	1	1/3	1/3	14%	3	0.23
G03 Data Interpretation	5	3	1	1/3	27%	2	0.44
G04 Consistency	9	3	3	1	53%	1	0.86

Six user stories are needed for implementing the *Terminology* priorities:

Figure 6-11: User Stories for Terminology

User Stories Topics	As a ... [functional user]	I want to ... [get something done]	such that ...[quality characteristic]	so that ... [value or benefit]
1) Q001 Audio	Train Operator	have an audio stream	I can transmit audio messages to passengers	they understand and are advised
2) Q002 Information	Train Operator	have freely programmable video information screens	I can transmit all necessary information to passengers	passengers are informed about connections and train statuses
3) Q003 Entertainment	Train Operator	use video screens for ads and news	passengers can follow the train ride	they know where they are and where they about to go
4) Q004 Train Status	Train Operator	see the status of all systems running the Double-Tiddlemutzz	I can perceive component failure early enough	train failure can be avoided
5) Q005 Terminology	Train Operator	address all components in their own language	each SW components receives the information it understands	I can combine part systems of various ages and generations
6) Q006 Training	Train Operator	train the terminology broker	communication improves over time	wear and tear can be combated

Functional Effectiveness is calculated the way same as before. Note the strong focus on *G04: Consistency*:

Figure 6-12: Functional Effectiveness for Terminology

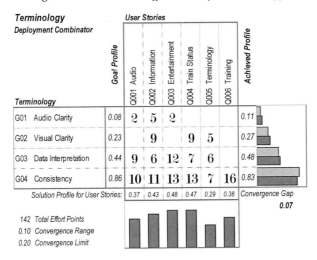

With 13 test stories, we can cover the six user stories for *Terminology* (Figure 6-13):

Figure 6-13: Initial Test Coverage for Terminology

User Stories	Goal Test Coverage	1) A.1 Direct Messages	2) A.2 Composed Messages	3) B.1 Switch Sources	4) B.2 Infotainment	5) C.1 Departure Table	6) C.2 Priority Messages	7) C.3 Learn about Priority	8) D.1 Train Location	9) D.2 Next Arrival	10) E.1 Standard Terms	11) E.2 Translation	12) F.1 Learn Terms	13) F.2 Use Learnings	Achieved Coverage
Q001 Audio	0.37	23	14	4	9	2	5	5	9	8	8	2	6	9	0.36
Q002 Information	0.43	10	15	6	4	9	11	10	9	9	14	14	2	7	0.43
Q003 Entertainment	0.48	14	9	16	15	19	11	10	16	11	6	12		8	0.52
Q004 Train Status	0.47	8	10	14	7	17	8	10	14	11	6	12		6	0.44
Q005 Terminology	0.29					4	6	10	2	2	16	14	13	15	0.29
Q006 Training	0.38	4	4	8	4	8		12	4	6	14	6	25	17	0.38
Ideal Profile for Test Stories:		0.31	0.28	0.27	0.22	0.33	0.22	0.29	0.30	0.25	0.31	0.31	0.20	0.30	Convergence Gap 0.05

688 Total Test Size
0.10 Convergence Range
0.20 Convergence Limit

Again, the details are left to the reader using the cloud data accompanying this book (Fehlmann, 2019).

6-3.5 THE HIERARCHY COMPARISON

The hierarchy comparison not only serves for connecting part comparisons but relies itself on software connecting the part solutions. It therefore has a data movement map describing functionality of its own that in turn must be effectively implement how parts interact. The data movements also play a role when extending initial test cases with ART; the test assertions travel along the data movements to extend test cases within existing test stories.

Figure 6-14: Pairwise Comparison for the Hierarchy

AHP Priorities / Operator's Needs	A ETCS	B Instrumentation	C Traction	D Electricity	E Comfort	F Doors	G Terminology	H Maintenance	Weight	Ranking	Profile
A ETCS	1	3	3	9	5	1	1/3	1/3	17%	2	0.47
B Instrumentation	1/3	1	1/3	1/5	1	3	3	1/9	10%	5	0.27
C Traction	1/3	3	1	1/3	1/3	1/5	3	1	9%	6	0.25
D Electricity	1/9	5	3	1	1/3	1/3	1	1/3	8%	8	0.22
E Comfort	1/5	1	3	3	1	1/3	1/3	1	8%	7	0.23
F Doors	1	1/3	5	3	3	1	1	1/3	13%	4	0.35
G Terminology	3	1/3	1/3	1	3	1	1	3	16%	3	0.42
H Maintenance	3	9	1	3	1	3	1/3	1	19%	1	0.50

The pairwise comparison (Figure 6-14) for the hierarchy comparison defines the operators' needs profile that serves as the goal profile for user stories describing the functionality of combining all the various D²TLDMUTS services into one train steering and control functional working place. Thus, we can again analyze and test this piece of software using the other applications as services – assuming already tested services.

The following user stories (Figure 6-15) implement these operators' needs:

Figure 6-15: Functional Effectiveness for Combining the Hierarchy

		As a ... [functional user]	I want to ... [get something done]	such that ...[quality characteristic]	so that ... [value or benefit]
1)	Q001 Traction	Train Operator	have the train running smoothly	all systems work together	energy consumption is minimized
2)	Q002 Comfort	Train Operator	to ensure convenient conditions for passengers	comfort is maintained	passengers feel well in the Tiddlemutzz
3)	Q003 Stop	Train Operator	make passengers exit and enter the Tiddlemutzz	exchange is fast	train stops can be kept short
4)	Q004 Monitor	Train Operator	know the wear & tear status of all components	failures can be prevented	maintenenace can be scheduled as needed

Figure 6-16: Functional Effectiveness for Combining the Hierarchy

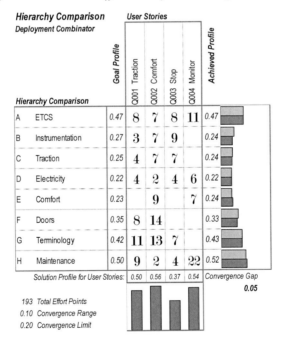

Figure 6-13 is a typical case of combining services. Many services are needed to fulfil basic functional needs.

Test coverage is calculated the same way as before (Figure 6-17):

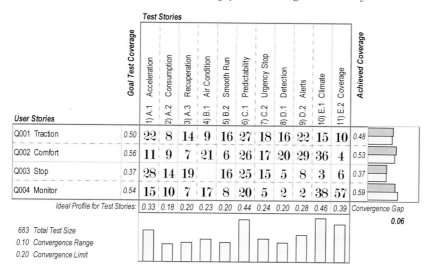

Figure 6-17: Initial Test Coverage for Combining the Hierarchy

User Stories	Goal Test Coverage	1) A.1 Acceleration	2) A.2 Consumption	3) A.3 Recuperation	4) B.1 Air Condition	5) B.2 Smooth Run	6) C.1 Predictability	7) C.2 Urgency Stop	8) D.1 Detection	9) D.2 Alerts	10) E.1 Climate	11) E.2 Coverage	Achieved Coverage
Q001 Traction	0.50	22	8	14	9	16	27	18	16	22	15	10	0.48
Q002 Comfort	0.56	11	9	7	21	6	26	17	20	29	36	4	0.53
Q003 Stop	0.37	28	14	19		16	25	15	5	8	3	6	0.37
Q004 Monitor	0.54	15	10	7	17	8	20	5	2	2	38	57	0.59
Ideal Profile for Test Stories:		0.33	0.18	0.20	0.23	0.20	0.44	0.24	0.20	0.28	0.46	0.39	Convergence Gap

Convergence Gap 0.06

683 Total Test Size
0.10 Convergence Range
0.20 Convergence Limit

Combining the previous test coverage matrices yields an initial test coverage matrix for the complete D²TLDMUTS (Figure 6-18).

6-3.6   THE FULL TEST COVERAGE MATRIX

Figure 6-19 in turn shows the test coverage matrix for the D²TLDMUTS after the first few automatic test case extensions. The empty spaces become filled with tests for the interactions between different apps.

To build the full test coverage matrix F, it is not good enough to add the sequence of test coverage matrices F_i, because the parts are of unequal importance for the customer. However, when multiplying each of the matrices E_i and F_i by the respective component of the solution profile $\bar{y}_i$ for the hierarchy comparison, the profiles remain the same and adding these matrices together yields a transfer function from all test stories into all user stories, thus the full coverage matrix.

$$F = \sum_{I=1}^{k} \bar{y}_i F_i, \qquad i = 1, \dots, k; k \in N; k > 0 \qquad (6\text{-}2)$$

Additionally, its convergence gap remains small because the convergence gaps of the part matrices were already small.

- 121 -

Figure 6-18: The D²TLDMUTS Initial Test Coverage Matrix, with the Combination App and the first two Part Apps – for Terminology and Door Control

User Stories	Goal Test Coverage	Test Stories: 1) A.1 Acceleration	2) A.2 Consumption	3) A.3 Recuperation	4) B.1 Air Condition	5) B.2 Smooth Run	6) C.1 Predictability	7) C.2 Urgency Stop	8) D.1 Detection	9) D.2 Alerts	10) E.1 Climate	11) E.2 Coverage
Q001 Traction	0.50	22	8	14	9	16	27	18	16	22	15	10
Q002 Comfort	0.56	11	9	7	21	6	26	17	20	29	36	4
Q003 Stop	0.37	28	14	19		16	25	15	5	8	3	6
Q004 Monitor	0.54	15	10	7	17	8	20	5	2	2	38	57
Q001 Audio	0.37											
Q002 Information	0.43											
Q003 Entertainment	0.48											
Q004 Train Status	0.47											
Q005 Terminology	0.29											
Q006 Training	0.38											
Q001 Stop	0.58											
Q002 Start	0.53											
Q003 Safety	0.35											
Q004 Pressure	0.50											
Ideal Profile for Test Stories:		0.33	0.18	0.20	0.23	0.20	0.44	0.24	0.20	0.28	0.46	0.39

User Stories	Test Stories: 1) A.1 Direct Messages	2) A.2 Composed Messages	3) B.1 Switch Sources	4) B.2 Infotainment	5) C.1 Departure Table	6) C.2 Priority Messages	7) C.3 Learn about Priority	8) D.1 Train Location	9) D.2 Next Arrival	10) E.1 Standard Terms	11) E.2 Translation	12) F.1 Learn Terms	13) F.2 Use Learnings
Q001 Traction													
Q002 Comfort													
Q003 Stop													
Q004 Monitor													
Q001 Audio	23	14	4	9	2	5	5	9	8	8	2	6	9
Q002 Information	10	15	6	4	9	11	10	9	9	14	14	2	7
Q003 Entertainment	14	9	16	15	19	11	10	16	11	6	12		8
Q004 Train Status	8	10	14	7	17	8	10	14	11	6	12		6
Q005 Terminology					4	6	10	2	2	16	14	13	15
Q006 Training	4	4	8	4	8		12	4	6	14	6	25	17
Q001 Stop													
Q002 Start													
Q003 Safety													
Q004 Pressure													
Ideal Profile for Test Stories:	0.31	0.28	0.27	0.22	0.33	0.22	0.29	0.30	0.25	0.31	0.31	0.20	0.30

User Stories	Test Stories: 1) A.1 Open Door	2) A.2 Pressure	3) B.1 Close Doors	4) B.2 Door occupied	5) B.3 Door blocks	6) B.4 All doors	7) C.1 Need Repair	8) C.2 Door unusable	Achieved Coverage
Q001 Traction									0.48
Q002 Comfort									0.53
Q003 Stop									0.37
Q004 Monitor									0.59
Q001 Audio									0.36
Q002 Information									0.43
Q003 Entertainment									0.52
Q004 Train Status									0.44
Q005 Terminology									0.29
Q006 Training									0.38
Q001 Stop	10		27	19	7	8	4	4	0.63
Q002 Start	7	12	19	17	7	4	5	5	0.54
Q003 Safety		6	6	3		10	19	17	0.31
Q004 Pressure	6	10	16	15	6	3	4	4	0.46
Ideal Profile for Test Stories:	0.23	0.23	0.66	0.52	0.20	0.21	0.23	0.22	

Figure 6-19: The Test Coverage Matrix, with the Combination App and the first two Part Apps – for Terminology and Door Control

Achieved Coverage (bar chart, by User Story): 0.52, 0.59, 0.37, 0.49, 0.32, 0.47, 0.45, 0.48, 0.39, 0.35, 0.58, 0.54, 0.41, 0.52

Test Stories – Block 1

User Stories	Goal Test Coverage	A.1 Acceleration	A.2 Consumption	A.3 Recuperation	B.1 Air Condition	B.2 Smooth Run	C.1 Predictability	C.2 Urgency Stop	D.1 Detection	D.2 Alerts	E.1 Climate	E.2 Coverage
Q001 Traction	0.50	22	8	14	9	16	27	18	16	22	15	10
Q002 Comfort	0.56	11	9	7	21	6	26	17	20	29	36	4
Q003 Stop	0.37	28	14	19		16	25	15	5	8	3	6
Q004 Monitor	0.54	15	10	7	17	8	20	5	2	2	38	57
Q001 Audio	0.37											
Q002 Information	0.43						8			11		
Q003 Entertainment	0.48	8				3						
Q004 Train Status	0.47							6				
Q005 Terminology	0.29				9						13	
Q006 Training	0.38					7						
Q001 Stop	0.58						11				3	
Q002 Start	0.53			2		3						
Q003 Safety	0.35							17				
Q004 Pressure	0.50	1					2			3		

Test Stories – Block 2

User Stories	A.1 Direct Messages	A.2 Composed Messages	B.1 Switch Sources	B.2 Infotainment	C.1 Departure Table	C.2 Priority Messages	C.3 Learn about Priority	D.1 Train Location	D.2 Next Arrival	E.1 Standard Terms	E.2 Translation	F.1 Learn Terms	F.2 Use Learnings
Q001 Traction	23	14	8								11		
Q002 Comfort						7						5	
Q003 Stop								1					
Q004 Monitor													
Q001 Audio	10	15	4	9	2	5	5	9	8	8	2	6	9
Q002 Information		9	6	4	9	11	10	9	11	14	14	2	7
Q003 Entertainment	14	10	16	15	19	11	11	16	9	6	12		8
Q004 Train Status	8		14	7	17	8	10	14	11	6	12	13	6
Q005 Terminology					4	6	6	2	2	16	14	25	15
Q006 Training	4	4	8	4	8		12	4	6	14	6		17
Q001 Stop			9			9							
Q002 Start										7			
Q003 Safety										6	21		
Q004 Pressure													

Test Stories – Block 3

User Stories	A.1 Open Door	A.2 Pressure	B.1 Close Doors	B.2 Door occupied	B.3 Door blocks	B.4 All doors	C.1 Need Repair	C.2 Door unusable
Q001 Traction				8			6	
Q002 Comfort		12					3	
Q003 Stop			5					
Q004 Monitor								
Q001 Audio								
Q002 Information		6		18			3	
Q003 Entertainment								
Q004 Train Status	11		8				8	
Q005 Terminology		21						
Q006 Training								
Q001 Stop	10		27	19	7	8	4	4
Q002 Start	7	12	19	17	7	4	5	5
Q003 Safety		6	6	3		10	19	17
Q004 Pressure	6	10	16	15	6	3	4	4

Ideal Profile for Test Stories:
Block 1: 0.33, 0.18, 0.20, 0.23, 0.20, 0.44, 0.24, 0.20, 0.28, 0.46, 0.39
Block 2: 0.31, 0.28, 0.27, 0.22, 0.33, 0.22, 0.29, 0.30, 0.25, 0.31, 0.20, 0.30
Block 3: 0.23, 0.23, 0.66, 0.52, 0.20, 0.21, 0.23, 0.22

Initially the full test coverage matrix F is sparsely filled: no test cases exist outside of the diagonal part matrices F_i. This means that no test cases cover the interactions between different part applications required by the A_1. However, such interactions exist and are essential for proper functioning of the whole complex system. Also, the initial set of test cases contains enough test stories that suggest test cases linking different part applications. In the D²TLDMUTS case, suitable test cases use the *Combination* application and eventually the Terminology application to move data across the other part applications.

Finding the relevant test cases for these test stories seems not difficult at all; except that there are quite a few. The real D²TLDMUTS has not only eight hierarchy levels but many more, and its part applications contain much more than just a few dozen functional size units. Consequently, the matrix becomes quite unhandy – for humans.

6-3.7   EXTEND THE TEST CASES

Not so for ART. Automatically extend the test cases in the white space requires nothing else than the application of the testing blockchain algorithm. The terminology application is paramount for combining test cases from various applications. Combining test cases from different parts of the diagonal also does the job, using combinatory logic.

The essence is that when selecting relevant test cases, the convergence gap must stay small while test intensity increases. The selection process depends from the effects on the convergence gap, just as with any other instance of ART.

This process of generating test cases and selecting those that keep the convergence gap small is not limited except by practical considerations how many tests eventually can be executed.

6-3.8   DO THE TESTS

Because the next step is executing the tests. If tests fail, fix defects found and re-execute the tests again with more test cases. These tests run on digital twins, if possible. It is not necessary to use a physical D²TLDMUTS; although a mockup would be helpful to test the wiring technology – which might be less that state of the art – and sometimes the networking technology is stone-age in real train systems.

Tests executed in the mockup can take as much time as needed; the 'real-time' adjective is optional now. Nevertheless, given that time is always precious, setting a time limit to extensive testing as still a valid idea. Testing can stop if no defects can be found anymore. There exist techniques that allow predicting the number defects not found yet; e.g., by using the exponentially weighted moving average as a sort of dynamically calculated control chart, proposed by Fehlmann & Kranich (Fehlmann & Kranich, 2014-1). Moreover, testing intensity – the average number of times a data movement is executed for testing – is another metric that allows determining when to stop testing.

6-3.9   Put these Complex Systems in Service

Now it is time to build these wonderful new railway cars – this needs a considerable amount of time, anyway – and try the complex new software-intense systems in the real world. There will be problems still, especially if the design was not excellent; however, these will rather not be software problems. There is reasonably good hope that ART already uncovered such problems and developers had time to fix it before the D²TLDMUTS goes into commissioning with the train operator.

6-4   Open Questions

One of the things that would be of high interest is knowing which cell values must be increased to close a convergence gap. Again, this refers to conducting a sensitivity analysis (see also section 5-2.5). This seems not impossible since the coefficients are linear and thus increments as well. However, the problem lies in the Eigenvector. It is well known that this kind of solutions have jumps; the primary eigenvector jumps from one position into the other. While small increments still might behave linearly, the jump from one principal eigenvector to another can happen anytime and is difficult to predict. The solution profile also has jumps and does not behave smoothly. Up to now we have no solution yet for this problem. For this reason, trying to identify the behavior of a certain cell is probably as hard as calculating the whole matrix. New research is underway to clarify that problem.

6-5   Conclusion

Extending tests by artificial intelligence becomes surprisingly simple once the underlying combinatory algebra is considered.

Note that these techniques can be applied even if little is known about how the part applications have been programmed. All that is really needed is a good investigation into what are the needs of the customer, e.g., the train operator.

Is this technique possibly useful for testing *Artificial Intelligence* (AI) itself? Remember, AI is basically a program whose algorithmic design is unknown; part of the training that the SVM received instead of the traditional programming.

CHAPTER 7: TESTING ARTIFICIAL INTELLIGENCE

Autonomous cars rely on visual recognition systems that use Artificial Intelligence (AI) for recognizing objects; for instance, an ADAS. They can be trained but they can also unlearn.

Testing image recognition systems requires creating new test images that can be used for Autonomous Real-time Testing (ART) of Advanced Driving Assistance Systems (ADAS) and autonomous vehicles. This is achieved with a data movement map according ISO/IEC 19761, serving as a model for image recognition.

7-1 INTRODUCTION

The death of Elaine Herzberg (August 2, 1968 – March 18, 2018) was the first recorded case of a pedestrian fatality involving an autonomous car, following a collision that occurred at around 10 PM Mountain Standard Time (UTC -7) in the evening of Sunday, March 18, 2018 (The National Transportation Safety Board, 2018). The following narrative is extracted from the said source.

Herzberg was pushing a bicycle across a four-lane road in Tempe, Arizona, United States, when she was struck by Volvo XC90 taxi outfitted with a sensor system, operated under test conditions by Uber. Since 2015, Uber conducted tests with various levels of automation in Arizona. The car was operating in self-drive mode with a human safety backup driver sitting in the driving seat. Following the collision, Herzberg was taken to the hospital where she died of her injuries.

According Uber, the accident was largely caused by the software that decides how the car should react to objects it detects. The car's sensors detected the pedestrian, who was crossing the street with a bicycle. Uber's software first registered Elaine Herzberg on lidar six seconds before the crash — at the speed it was traveling, that puts first contact at about 115 m away. As the vehicle and pedestrian paths converged, the self-driving system software classified the pedestrian first as an unknown object, then as a vehicle, and then as a bicycle with varying expectations of future travel path. The software decided it did not need to react right away. Like other autonomous vehicle systems, Uber's software can ignore "false positives," or objects in its path that are not an obstacle for the vehicle, such as a plastic bag floating over a road.

Then, 1.3 seconds before impact, which is to say about 24 m away, the self-driving system determined that an emergency braking maneuver was needed to mitigate a collision. According to Uber, emergency braking maneuvers are not enabled while the vehicle is under computer control, to reduce the potential for erratic vehicle behavior. The vehicle operator is relied on to intervene and act. The system is not designed to alert the operator. The Volvo model's built-in safety systems — collision avoidance and emergency braking, among other things —were also disabled while in autonomous testing mode.

The self-driving system data showed that the vehicle operator intervened less than a second before impact by engaging the steering wheel. The vehicle speed at impact was 62 km/h. The operator began braking less than a second after the impact. The data also showed that all aspects of the self-driving system were operating normally at the time of the crash, and that there were no faults or diagnostic messages.

The dead of Elaine Herzberg raises one major question: Why were the visual recognition systems tested in real life situations, instead of under labor conditions?

7-2 HOW TO TEST ARTIFICIAL INTELLIGENCE

Computer Vision and *Artificial Intelligence* (AI) overlap. AI is different from ordinary software by its capability to learn. This means, AI can adapt to new environments, data, images and videos. While AI can be used for other tasks, computer vision is concerned with the theory behind artificial systems, extracting information from images. Areas of AI deal with autonomous planning or deliberation for robotical systems to navigate through an environment. A detailed understanding of these environments is required to navigate through them. Information about the environment could be provided by a computer vision system, acting as a vision sensor and providing high-level information about the environment and the robot.

AI and computer vision share other topics such as pattern recognition and learning techniques. Consequently, computer vision is sometimes seen as a part of the AI field. Testing AI in computer vision obviously is not so straightforward; mainly, because it is not possible to predict what is the correct outcome. The test case might produce different responses, and all are correct at a given state of experience collection.

Recall that AI basically is sorting data into categories based on previous learning, or sample sets. The Uber car did exactly that when its Lidar, and ten visual cameras, recognized the object moving towards the car's driveway (The National Transportation Safety Board, 2018). The difficulty was to find the right category. Humans encounter the same difficulty, when a biker enters the road from the pedestrian sidewalk. Expecting a pedestrian, they rapidly must adapt categories to a bicycle that moves differently and follows different traffic rules than a pedestrian. Things become even more complicated if suddenly the pedestrian conjures up a skateboard, or a scooter. Traffic rules for the latter two conveyances are unknown, or do not exist. Humans are disturbed, and so are visual recognition systems.

Since the important contribution of the visual recognition system is categorization, it should be tested whether categories detected by the visual recognition system remain the same over its lifetime. But that is not enough. Behavior on certain sample image sequences should also remain stable – except if new learnings tell it otherwise. Obviously, tests must adapt to learnings. On the other hand, learning systems can become neurotically disturbed – sick, like humans (van Gerven & Bothe, 2018). Thus, this is a case for *Autonomous Real-time Testing* (ART). For using AI in safety-critical environments, testing AI is required anytime, autonomous, without human intervention.

7-2.1   BASELINING

You start testing AI as any other software

- Identify the software under test
- Identify the goals of testing
- Draw a data movement map that explains the user's view on its functionality
- Calculate functional effectiveness to make sure it does what users expect
- Adjust scope of testing until goal and functional effectiveness converge
- Prepare the test stories:
    o Identify new test stories
    o Fill test stories by test cases
    o Calculate test coverage
- Repeat above three steps until test coverage converges
- For each test story, generate more test cases:
    o Apply the test case variation rules defined in Table 5-3
    o Thus, generating even more test cases
- Repeat generating more test cases per test story until test coverage converges

Perform the tests and validate test stories and test cases. Identify defects and remove them, or mitigate them, until your system is defect-free.

7-2.2   EXTENDING TEST CASES

Use the algorithm explained in section *5-2: Generating New Test Cases* to expand the test suite. Consider the AI domain when expanding the testing blockchain. For instance, for traffic vehicles, use video sequences form traffic scenes to add to new test cases. Use video sequences that have been used for deep learning and other who were not. You must manually classify the videos for the category of traffic it represents; it is therefore the same kind of work for testing as for learning.

As always with ART, you keep the test stories from the initial test suite stable while adding more test cases to improve test intensity and to detect more defects. For visual systems, the primary source for new test cases are new images and videos.

Keeping test coverage good enough is somewhat easier than in other ART instances, since you only exchange test data. You do not change the aim of testing; not even incrementally.

7-2.3   INTERPRETING TEST RESULTS

In fact, it does not matter if you take all learning videos for testing or not. It is unlikely that you get a higher degree of trust in your AI system whether you show him all tests in advance. Unlike humans, who might remember learning videos but need extra effort to verify their learning, machine intelligence always can recall what they once have seen before; but the question is whether they still put those videos in the same categories as in the beginning.

The aim of AI testing is to verify stable behavior in categorization as previously learned. This is different from human learning where humans should be able to interfere correct evaluations from their skills. As already mentioned, there is nothing intelligent with AI. Testing machine intelligence means verifying that the software keeps identifying the same categories and does not change them. Testing AI remains simple while no new categories are added.

If something else is being tested than categorization, interpreting test results can become quite difficult. Remember that test results should be known in advance. AI behavior is not known before.

Evaluating test results is therefore a manual task, supported by AI but delegating responsibility back to the human in case the response of the test case is something else than one of AI's established categories.

Adding another category to AI is connected to re-learning from scratch. You must supply all given evidence again and accept that the category borders move. In such cases, testing AI also starts from the beginning with establishing a new baseline.

7-2.4   NEVER STOP TESTING – REPEAT TESTING FOREVER

Not only learning data changes, categories themselves are not except from change. Certain categories such as legal behavior in traffic are also subject to change and must be adapted to new environments and facts. Testing AI will detect such changes.

Therefore, for the lifetime of the AI system, testing must repeat. AI systems consist not of stable, always repeatable software but depend from their environment. If the AI system fails to reproduce correct answers, it might indicate a shift in the learning data and probably learning must restart from the beginning. Such restarts are typically required, for instance in traffic, if new conveyors appear, such as scooters, electro-scooters, electro-bikes, and if rules change, for instance if fast electro-bikes are no longer admitted on cycle paths.

Testing AI happens typically if the AI system is idle. Only in rare cases a test that interrupts and competes with actual operations might be useful, for instance when encountering unexpectantly a new environment. If a car unexpectedly meets local traffic that is typical for urban areas, and the car believes it is overland, then it might indicate the need for retesting the map services used. When map service problems can be excluded, the car might run through a newly developed housing area - or a squatter habitation – and inform its map services about this. The map service can then decide from this and similar notifications whether it needs adjourning the map.

7-2.5   LOCALIZATION

There are also other geographical factors. For instance, in certain countries a pedestrian moving towards a pedestrian crossing causes car traffic to stop. Pedestrians have priority. In certain other countries, if you stop your car to let a pedestrian strip, you risk a rear-end collision. Other road users would be surprised. Such differences in the practices adopted in road traffic can exist despite quite similar road traffic regulations.

This makes ART not simpler. To use the same test suite for different locations involves the risk that such local practices are not reflected. In such cases, an autonomous car that "learned" driving in one country is not easily acceptable on other roads.

7-2.6   WHEN TO TEST ARTIFICIAL INTELLIGENCE?

We already mentioned that AI must be testable "anytime". Nevertheless, no system is anytime available for testing. The typical times a system does ART are when idle. Since idling can be stopped anytime, running test must also be able to stop immediately. This is possibly not so easy if sensors and actuators are involved that first need being reset before use. On a *Digital Twin*, such tests run smoother.

Users of AI systems therefore must be able to see when their system is running tests. It is also recommendable that users see results of tests. Section *1-4.2: Consumer Metrics* proposes a standard how to represent test results for consumers. It is obvious that such representations are complimentary to the full test suite records that are probably of more interest to the system supplier than to the consumer.

7-3 A Deep Learning Application as a Sample

We take our first example from *Chapter 4: Testing Privacy Protection and Safety Risks* and use it now to demonstrate how to test the *Look & Act in ADAS* as shown in Figure 4-4.

7-3.1 XAI – Explainable Artificial Intelligence

However, we must go deeper into the details without really knowing how the *Visual Recognition System* (VRS) works. Interestingly, we do not need to know how the VRS was implemented. It does not matter whether the VRS uses programmed algorithms or whether a neural network has learned to behave correctly.

An automatic generation of the data movement map is not possible without code. But we can draw a data movement map that delivers what we want, using our understanding of the VRS. The ISO/IEC 19761 standard and the data movements maps enable software measurements without code.

Explainable AI addresses this problem – how can you understand and comprehend decisions of an AI-enabled device that probably used deep learning to learn correct decisions? Such devices are now omnipresent and gradually replacing older decision algorithms that proved considerably less reliable but have code that can be assessed and eventually understood. Nevertheless, regulators ask for explanations.

Theodorou provides a robust definition of transparency as a mechanism to expose the decision making of a robot (Theodorou, et al., 2017). Since 2017, the *Defense Advanced Research Projects Agency* (DARPA) conducts a project providing explainable decision models and enable humans to understand, appropriately trust, and effectively manage the emerging generation of artificially intelligent partners. For additional information, see Gunning (Gunning, 2017).

A data movement map explains any AI device consistently and effectively, the convergence gap of the test coverage transfers function guarantees relevance. Regulators would better ask for test coverage than whatever an AI device may produce as "explanation". Remember that it is quite easy for an AI system to learn what kind of explanations humans accept. Whether those explanations guarantee correct decisions is not part of the question asked.

7-3.2 THE GOAL OF TESTING

As before, we need the goal profile to do testing. These goals are not the same as the car users' needs for ADAS, and obviously we need dedicated user stories. First, the user of the visual recognition system is not the car user, but the car itself, represented by the car's ADAS. Second, we focus on the visual recognition system and how it visually understands and interprets the environment using its cameras and Lidar. For modeling this different viewpoint on testing (or explaining) the VRS, we clearly need more data movements and thus we need to look deeper into the VRS app in Figure 4-4: *Look & Act in ADAS*.

Although this is an arbitrary viewpoint, we expect that three levels of decisions are taken, and monitored, when executing the VRS app:

- A top-level decision: Is the object hard, soft, or possible a blur only? Sometimes, fogs look like a cat, empty plastic bags simply fly around.

- The next level is whether the object moves actively, or passive, or not at all. This requires sensing wind, rain and other weather events.

- The third level is assigning it a traffic category such as pedestrian, bike, other car, or fixed installations such as a signal, a post, or curbstone.

With our preferred method for prioritization, the pairwise comparison or simple AHP, we get the following profile (Figure 7-1):

Figure 7-1: The Visual Needs

Customer's Needs Topics	Attributes			Weight	Profile	
				AHP Priorities		
y1 Recognize Objects	Distinguish from background	Movable	Rolling or not	**13%**	0.32	
y2 Impact Category	Hard	Soft		**18%**	0.46	
y3 Reaction Category	Active	Passive	None at all	**13%**	0.32	
y4 Traffic Category	Cars	Bikes	Pedestrians	**10%**	0.25	
y5 Movement	Direction	Speed	Variability	**22%**	0.57	
y6 Blur Resilience	Minimum outline	Fog	Snow or rain	**8%**	0.19	
y8 Distance	Lidar measurements			**16%**	0.40	

The decisions originate from the following AHP (Figure 7-2):

Figure 7-2: The Visual Needs Priority AHP

AHP Priorities / Customer's Needs	y1 Recognize Objects	y2 Impact Category	y3 Reaction Category	y4 Traffic Category	y5 Movement	y6 Blur Resilience	y8 Distance	0.18	0.13	0.10	Weight	Ranking	Profile
y1 Recognize Objects	1	1	1/3	1/3	1/9	1/3	9	0.20	0.05	0.03	**13%**	4	0.32
y2 Impact Category	1	1	3	3	1	1	3	0.20	0.45	0.24	**18%**	2	0.46
y3 Reaction Category	3	1/3	1	2	1	3	2	0.07	0.15	0.16	**13%**	5	0.32
y4 Traffic Category	3	1/3	1/2	1	1/3	6	1/3	0.07	0.08	0.08	**10%**	6	0.25
y5 Movement	9	1	1	3	1	9	1/6	0.20	0.15	0.24	**22%**	1	0.57
y6 Blur Resilience	3	1	1/3	1/6	1/9	1	1	0.20	0.05	0.01	**8%**	7	0.19
y8 Distance	1/9	1/3	1/2	3	6	1	1	0.07	0.08	0.24	**16%**	3	0.40

It is not surprising that *y5: Movement* and *y2: Impact Category* are highest in ranking. These are the most important visual needs for driving a car. On the other hand, *y6: Blur Resilience*, the ability to recognize objects even in fog or precipitation, is a precondition for the others, but by itself it is not dominant.

Consequently, an ADAS needs a Lidar; otherwise, recognizing objects and thus movements and impact category is difficult to achieve if cameras only rely on signals in the visible range. As usual, already the AHP points at the relevant technical challenges.

7-3.3 USER STORIES FOR THE VRS

The application modeled in Figure 7-4: *Data Movement Map for the Visual Recognition System (VRS)* implements the following eight user stories (Table 7-3):

Table 7-3: Visual Recognition User Stories

Label	As a ...	I want to ...	Such that ...	So that ...
Identify Objects	Car ADAS	understand objects around me	I do not hit any of them	I can have a smooth drive
Identify Movements	Car ADAS	understand which objects move and where they move	I can calculate my free way	I can have a smooth drive
Identify Dangers	Car ADAS	distinguish objects from background environment	I get no false alarms	I do not stop unnecessarily
Predict Reactions	Car ADAS	understand whether an object moves actively or passively	I can predict where it's moving	I can adapt my route
Identify Traffic	Car ADAS	identify traffic participants	I can predict their speed	I can adapt my route
Collect Images	Car ADAS	extract relevant information from images	I understand my environment	I can use experiences for later learning
Blur Independence	Car ADAS	have vision despite fog and precipitation	I can drive despite limited visibility	bad weather does not stop me
Plausibility	Car ADAS	be sure the VRS returns a valid object catalog	I can rely on its findings	I won't get disturbed

The data movement map in Figure 7-4 on the following page implements these user stories.

Figure 7-4: Data Movement Map for the Visual Recognition System (VRS)

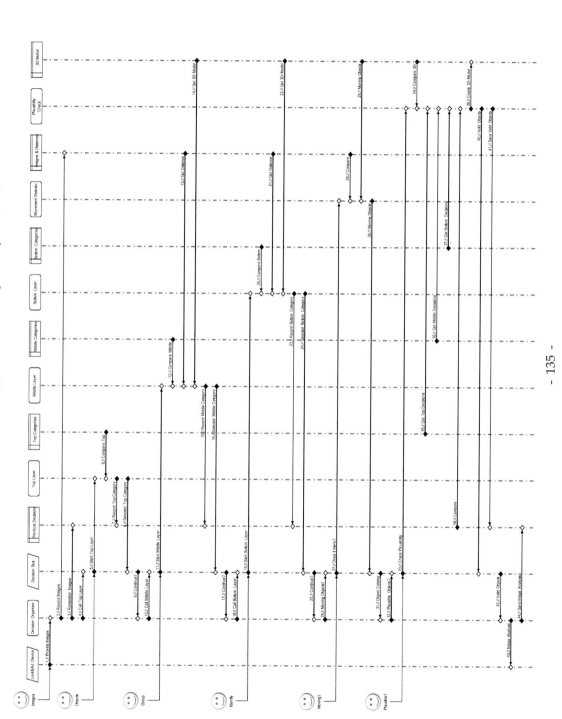

Deploying the user stories against the visual needs yields the transfer function shown below in Figure 7-6. Not surprisingly, *Q004: Predict Reactions* is the most important of our eight short user stories.

Figure 7-5: User Story Priority

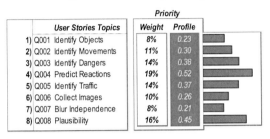

	User Stories Topics	Priority	
		Weight	Profile
1) Q001	Identify Objects	8%	0.23
2) Q002	Identify Movements	11%	0.30
3) Q003	Identify Dangers	14%	0.38
4) Q004	Predict Reactions	19%	0.52
5) Q005	Identify Traffic	14%	0.37
6) Q006	Collect Images	10%	0.26
7) Q007	Blur Independence	8%	0.21
8) Q008	Plausibility	16%	0.45

Remember that we had no clue how our VRS determines the list of valid objects that it recognizes. Possibly a *Support Vector Machine* (SVM) is used; see Gunn (Gunn, 1998), and more recently Pupale (Pupale, 2018). However, we use our data movement map model from Figure 7-4 to assess functional effectiveness with the later goal of testing.

Figure 7-6: Functional Efficiency – User Story Deployment based on Figure 7-4

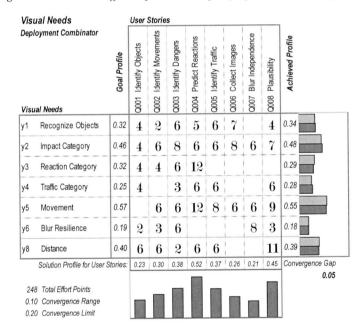

Visual Needs Deployment Combinator	Goal Profile	Q001 Identify Objects	Q002 Identify Movements	Q003 Identify Dangers	Q004 Predict Reactions	Q005 Identify Traffic	Q006 Collect Images	Q007 Blur Independence	Q008 Plausibility	Achieved Profile
y1 Recognize Objects	0.32	4	2	6	5	6	7		4	0.34
y2 Impact Category	0.46	4	6	8	6	6	8	6	7	0.48
y3 Reaction Category	0.32	4	4	6	12					0.29
y4 Traffic Category	0.25	4		3	6	6			6	0.28
y5 Movement	0.57		6	6	12	8	6	6	9	0.55
y6 Blur Resilience	0.19	2	3	6				8	3	0.18
y8 Distance	0.40	6	6	2	6	6			11	0.39
Solution Profile for User Stories:		0.23	0.30	0.38	0.52	0.37	0.26	0.21	0.45	Convergence Gap 0.05

248 Total Effort Points
0.10 Convergence Range
0.20 Convergence Limit

There is a clear focus on predicting reactions and check distances for plausibility in the data movement map. This is what we expect from a VRS but do not know how it is implemented by the SVM or any other neural network. The functional effectiveness matrix identifies the data movements that implement a specific user story.

Test Coverage is calculated from the following fourteen test stories:

Figure 7-7: Test Stories with two Test Cases

		Test Story	Case 1	Test Data	Expected Response	Case 2	Test Data	Expected Response
1)	A Objects	A.1 Object Contour	A.1.1	Object; Background	Contour exact	A.1.2	Object; Fog; Background	Contour somehow
2)		A.2 Object Move	A.2.1	Object; Move active	Move Vector	A.2.2	Object; Move passive	Move Vector
3)	B Prediction	B.1 Predict Move	B.1.1	Object; Move; Identity	Move Vector	B.1.2	Object; Move; Unknown	Move Range
4)		B.2 Predict Collision	B.2.1	Object; Move Vector; Identity	Collision Point	B.2.2	Object; Move Vector; Unknown	Collision Range
5)		B.3 Predict Reaction	B.3.1	Identity	Move Vector	B.3.2	Object; Move Vector; Unknown	Action Range
6)	C Identification	C.1 Identify People	C.1.1	Pedestrian; Walking	Move Vector	C.1.2	Pedestrian; Stagnant	Action Range
7)		C.2 Identify Child	C.2.1	Child; Playing	Collision Range	C.2.2	Child; Watching	Action Range
8)		C.3 Identify Car	C.3.1	Car; Move Vector	Collision Range	C.3.2	Car; Braking slow	Collision Range
9)		C.4 Identify Truck	C.4.1	Truck, Move Vector	Collision Range	C.4.2	Truck; Braking slow	Collision Range
10)		C.5 Identify Bike	C.5.1	Bike; Move Vector	Collision Range	C.5.2	Bike; Stopping	Collision Range
11)		C.6 Identify Blur	C.6.1	Object; Blur; Move Vector	Identify	C.6.2	Blur; no object	Identify
12)		C.7 Identify Position	C.7.1	Objects; Identified; Move Vectors	Move Model	C.7.2	Objects; Identified; Stagnant	Position Model
13)	D 3D-Model	D.1 Use 3D-Model	D.1.1	3D Position, Identified, Move Vector	Move Model	D.1.2	3D-Model; Identified; Stagnant	Position Model
14)		D.2 Verify 3D-Model	D.2.1	Move Model, Move Vector	3D-Position	D.2.2	Objects, Stagnant	Position Model

There are many more than two test cases per test story: however, not shown here. Based on this, we get the following test coverage (Figure 7-8):

Figure 7-8: Baseline Test Coverage

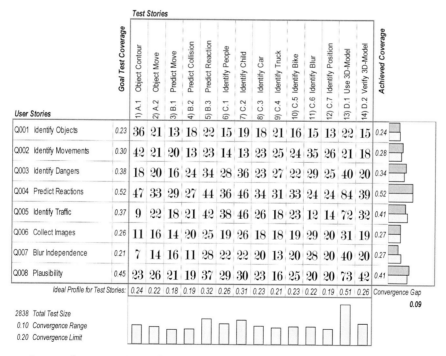

The main test focus receives *Q004: Predict Reactions*; as expected. This can be seen when enhancing the highest frequency cell by color, or bold type, display. Also, user story *Q008: Plausibility* receives support by all tests; this is because results always flow into the decision repository fueling later plausibility checks.

The need for the test stories *D.1: Use 3D-Model* and *D.2: Verify 3D-Model* became apparent after it proved impossible to achieve a convergence gap below 0.10 (10%) with only the twelve test stories directly addressing ADAS functionality. Thus, the assumption of the tester, that the VRS uses a kind of three-dimensional model to take informed decisions, is supported by the ART testing algorithm. Whether the "intelligence" inside the VRS does it this way, or another way, remains open but is irrelevant.

The total test size statistics looks as follows:

Figure 7-9: Baseline Test Status Summary

Total CFP:	44	Test Size in CFP:	2838
		Test Intensity:	64.5
Defects Found in Total:	0	Defect Density:	0.0%
Defects Pending for Removal:	0	Data Movements Covered:	100%

An initial test intensity of 64.5 is not bad; it looks we have mapped enough data movements for the necessary granularity. This allows explaining and testing the expected qualities of the VRS, including plausibility checks and categorization of the various traffic participants.

At least, we have an idea how to test a VRS before it hits the roads.

7-4 NEXT STEPS, AND A PRELIMINARY CONCLUSION

Clearly, a visual system needs more tests than those shown in this chapter. We use ART to generate more test cases out of the fourteen test stories to increase test intensity. However, at the current stage of research, we have no clue what test intensity is enough for a VRS in an autonomous car.

Applying ART means adding more test cases, more image sequences, always with respect to the convergence gap, aiming at improving it towards less than 0.1. The convergence gap of less than 0.1 indicates that the current test suite misses the goal profile by less than 10% (Fehlmann, 2016, pp. 13,31). This limits combinatorial explosion, as it allows selecting relevant test cases only.

The basic idea how to deal with "untestable" neuronal networks and deep learning SVMs is to create a model. This model describes what we think how it should work, and then we use *API Test Automation* (Reichert, 2015) to ask the right questions to the intelligent device, as indicated in the test cases. Obviously, this requires the ability of the device to answer more questions that those primarily intended.

7-5 A SIDE NOTE

SVM, Perceptron, Combinatory Logic are all inventions of the first and second third of the 20th century. The original SVM had been described by Vladimir Vapnik at times, where he possible never had the opportunity to touch a computer except huge mainframes without nominal computing power. The original combinatory logic algorithm for generating new test cases – or formulas about tests – has been implemented in 1980 on a DEC-10 at the Center for Interactive Computing of the ETH Zurich, by the author (Fehlmann, 1981).

There is nothing new about AI; it had been rediscovered and put to work because finally computing power is available almost for free. And again, there is nothing intelligent about AI. It is all about searching big data, and classifying vectors describing objects of the real world.

Preparing the reference vectors for deep learning is hard work by intelligent, insightful people. The same is true when preparing test stories and initial test cases for testing AI. The rest of the work is ephemeral: big calculations with much data elaborating on the rationale of skilled humans.

However, the real beauty of all these stories is: all the ingredients are well-known, only needing rediscovery. We only had to put old threads in a new way together.

CHAPTER 8: AGILE TESTING WITH THE BUGLIONE-TRUDEL MATRIX

While functional effectiveness is enough to calculate test coverage automatically, focusing in development on functional effectiveness alone does not guarantee developing a product continuously towards increased customer satisfaction. Customers might have other requirements than functionality alone. Developers thus need to keep an eye on both, functional effectiveness and on-functional customer needs.

This chapter describes modern software development that harvests on the teams' experience and expertise to continuously provide world-class customer experience using the Buglione-Trudel Matrix introduced in the "Managing Complexity" book of 2016.

Autonomous real-time testing is also useful for continuously observing and measuring customer experience.

8-1 INTRODUCTION

Readers of the previous book (Fehlmann, 2016, p. 200ff) remember how the *Buglione-Trudel* (BT) *Matrix* helps agile teams to organize themselves, elicit requirements and adapt easily to changed goals in product development. The crucial point is to use the value seen by customers also in a transfer function that is possibly not equal to the functional effectiveness. The BT matrix complements functional effectiveness by taking non-functional requirements into account and develop work along the lines of value perceived by the customer. The customer and its values are typically represented by the *Product Owner*. The BT matrix is basically an interactive story board for agile teams where the story cards are valuated against the customer's needs. The *Story Cards* represent those parts of a user story that is selected within a given sprint. Often, a user story splits into a functional story card, implementing functionality, and one to several non-functional story cards, implementing quality characteristics that also need time and effort.

For this, we distinguish the *Sundeck* and the *Cellar* of the BT matrix; see Figure 8-1. Both are transfer functions, mapping user stories, respectively its story cards, to customer needs. However, the sundeck is interactively designed by the development team, while the cellar remains much more stable. The functional requirements in general are more stable and less influenced by the development team than the quality, or non-functional, requirements. However, all is subject to new learnings and change of environment; the transfer functions adapt themselves and possible need rework to keep the convergence gap low. For more information and sample BT matrices, see the previous book (Fehlmann, 2016, p. 201).

Figure 8-1: Deming Chain for Agile Software Development and Software Testing

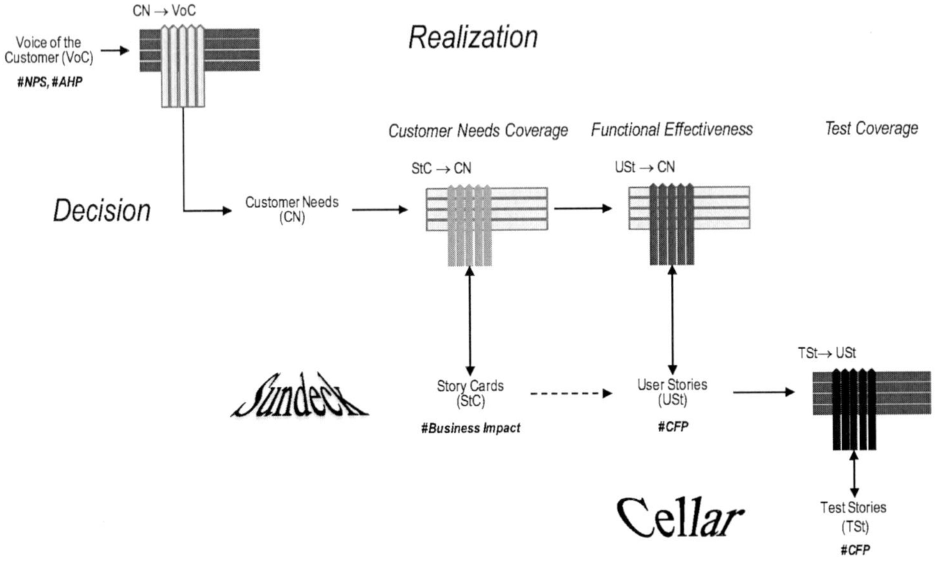

Formally, the *Customer Needs Coverage* and the *Functional Effectiveness* transfer function look similar, since both rely on user stories. However, since the first uses the valuations of the customer and the second the functional size for the matrix cells, its results can diverge, and most often they do.

Since automatic testing is possible for functionality only, we can rely on the functional effectiveness for assessing test coverage, while some of the quality aspects – if not linked to any functionality – cannot be tested automatically. Whether some output screen is readable and convenient to users, looking attractive, only users can tell, unless we can test its layout against certain ergonomic rules and regulations.

On the other hand, development must follow the customer's priorities and provide value. Such value can be other than functionality; for instance, does adherence to corporate identity rules provide high value but no functionality.

8-2 STORY CARDS WITH TEST STORIES

To deal with these constraints and requirements, we use functional and non-functional story cards, plus a mix of both. Story cards can have functionality to implement, or simply call for adding value by adding quality features that affect things such as ease-of-use, or appearance, or information presentation to humans, even to machine users.

Thus, a story card describes one of the tasks needed to implement a user story, references the functionality affected by such work, ideally as a data movement map, and identifies the business value of this task. Thus, it might be that the business value does not so much originate from the functionality but from other aspects such as establishing credibility and trust among users of a software.

We use the ADAS example from section 4-3 and distribute its user stories on four sprints. This yields a story card table as shown in Figure 8-2:

Figure 8-2: Story Table for ADAS

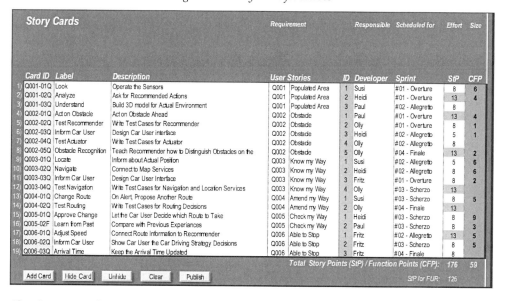

For instance, the user story *Q001: Populated Area* is implemented with three story cards, spanning over two sprints:

- *Q001-01Q Look*   Operate the Sensors (StP: 8; CFP: 6)
- *Q001-02Q Analyze*   Ask for Recommended Actions (StP: 13; CFP: 4)
- *Q001-03Q Understand* Build 3D model for Actual Environment (StP: 8; no CFP)

Looking at the three story cards that implement user story *Q001: Populated Area*, we see one (Figure 8-3) that implements main functionality by accessing the sensors and collecting data from them. The next story card (Figure 8-4) analyzes the situation and provides recommendations for the ADAS. This story card has highest business value as this is what the car user expects from the ADAS. The third story card is about how the functional process *F001: Car Driving Function* asks for action. To do this, it creates a 3D model of the actual road situation, with predictions what the other vehicles and people on the road are likely to do next, that it can submit to the *A003: Recommender*.

Figure 8-3: Q001-01Q: Look - Operate the Sensors

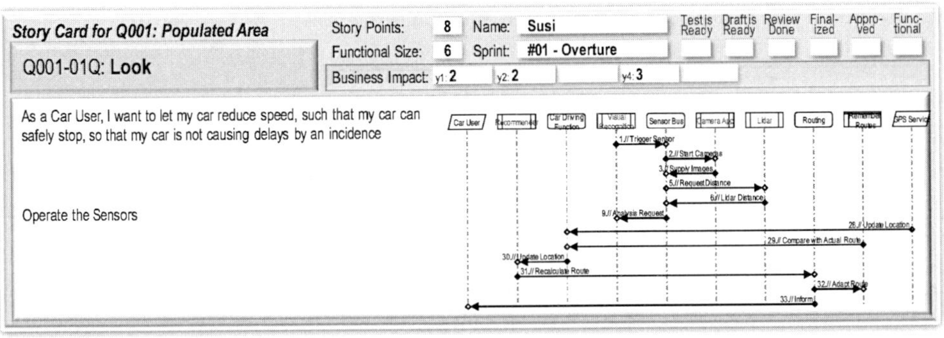

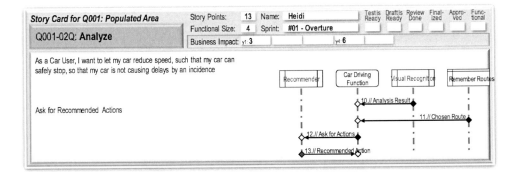

Figure 8-4: Q001-02Q: Analyze - Ask for Recommended Actions

The third story card has no extra functionality, since constructing the 3D model is contained in the functional process *F001: Car Driving Function*. It does not require extra data movements – except if we change focus and granularity and ask how the various functional users in the functional process *F001: Car Driving Function* perceive the steps needed for car driving.

Thus, the last story card shown in Figure 8-5 provides no new functionality in terms of data movements but implements the algorithm needed to let the VRS make a recommendation. It is left open whether the mentioned 3D model is built by an algorithm or learned by an SVM or neural network. In any case, the external application *A001:*

Visual Recognition (VRS) provides a suitable 3D model that can be used to make recommendations for steering and acting by the ADAS. The actual recommendation originates from another external application *A003: Recommender*.

Figure 8-5: Q001-03Q: Understand - Build 3D model for Actual Environment

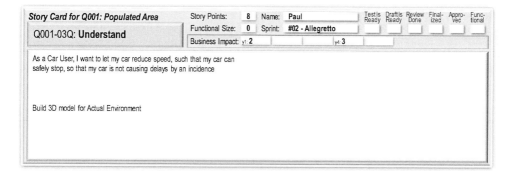

8-3 SELECTING TEST STORIES FOR STORY CARDS

The back of the story cards contains the applicable test stories. Since the user story *Q001: Populated Area* is quite prominent for ADAS functionality, ten of the eleven test stories (see section *4-3: ART for ADAS*) are listed. Thus, the developers know against which test stories their functionality will be tested. Test stories clarify requirements. However, the aim is two ways: the developers are encouraged to write additional test cases that they think relevant.

The selection of test stories on the back of story cards is automatic: all test stories that contain a test case testing one of the data movements occurring in the user story are listed. Thus, the story card *Q001-03Q: Understand - Build 3D model for Actual Environment* also features all ten test stories that affect user story *Q001: Populated Area* even if the story card is not referring directly to any data movement. Nevertheless, it might make sense since testing any of the non-functional quality characteristics involves some functionality – otherwise, it would not be a test, rather a static assessment.

As you always need functionality to implement non-functional characteristics, you always need functional tests for testing quality characteristics.

8-4 Creating Test Stories by the Development Team

Since our cards use "intelligent paper" – i.e., they are distributed and available electronically – adding test cases is a matter of harvesting developers intelligence for creating relevant tests. Thus, the test suite grows while the product evolves.

Obviously, tests run as soon as enough functionality is available. This is the same kind of automated test runs that is usually in place for unit tests delivering the "green bar" needed for the daily build.

Harvesting skills and intelligence of the development team – this is the reason why we institutionalize collection of test stories and test cases while developers look at the details of implementing user stories.

The back of the story cards is not immutable but is used to collect test stories and test cases. Initially, when development starts, the test stories might even be missing, and it is up to the development team to propose them. If every developer proposes tests, it needs a *Test Manager* who collects these proposals, identifies when the same test story is proposed twice, or a test case is assigned to the wrong test story. The test manager arranges the back of the story cards.

8-5 Test Management

Test management is probably the most important task in ART. That testing starts at the beginning of any product development, is already clear. Most agile software development uses *Test-Driven Development* (TDD) (Poppendieck & Poppendieck, 2007), as already mentioned. ART extends TDD based on existing test cases and can be used to increase test intensity already while developing the product.

Setting up the test stories goes in parallel with the user stories and controlling evolution of knowledge all through the development stages and sprints by functional effectiveness and test coverage transfer functions starts at the very beginning of product development. Especially, if the product is complex or safety critical.

8-6 Conclusions

Thanks to the ongoing controlling of convergence gaps in all transfer functions involved, developing software even for safety critical application such as automatic driving, for artificial intelligence, or for complex software-intense systems becomes feasible.

Technology advances cannot become successful without developing suitable control mechanisms as institutionalized with software testing. The dream of autonomous vehicles seems nowadays, by mid-2019, remaining a dream. Whether ART alone can put the dream into reality is not sure. ART detects defects and avoids fatal failures but does not solve the problem how to drive through Naples or Delhi. There human-to-human communication between car drivers is much more important than sensors and car-to-car communication.

Nevertheless, the future is with software-intense systems; but the future still lacks ART. Developing tools for ART is probably right now the most urgent task for the ICT community.

While it is not sure whether ART helps avoiding catastrophic failures, ART creates an open space, the combinatory algebra of arrow terms, where unthinkable test cases have a well-defined place. While AI, as already stated, is not intelligent, AI can help people to think much farther than ever and anticipate consequences of their new technology that they are going to develop and impose on society.

BIBLIOGRAPHY

Akao, Y., ed., 1990. *Quality Function Deployment - Integrating Customer Requirements into Product Design.* Portland, OR: Productivity Press.

Barwise, J. et al., 1977. *Handbook of Mathematical Logic.* Studies in Logic and the Foundations of Mathematics ed. Amsterdam, NL: North-Holland Publishing Company.

Bell, D., 2004. *UML basics: The Sequence Diagram – Introductory Level.* [Online] Available at:
http://www.ibm.com/developerworks/rational/library/3101.html
[Accessed 27 Nov. 2019].

Cabré Castellví, T. et al., 2017. *El multilingüisme en blanc i negre.* 1 ed. Barcelona: Càtedra Pompeu Fabra-Universitat Pompeu Fabra.

Cagley, T., 2018. *Using Size to Drive Testing in Agile.* s.l.:Webinar - Verbal Communication.

Cairns, H., 2014. *A short proof of Perron's theorem.* [Online] Available at: http://www.math.cornell.edu/~web6720/Perron-Frobenius_Hannah%20Cairns.pdf
[Accessed 25 August 2015].

COSMIC Measurement Practices Committee, 2017. *The COSMIC Functional Size Measurement Method – Version 4.0.2 – Measurement Manual,* Montréal: The COSMIC Consortium.

Ebert, C. & Jones, C., 2009. Embedded Software: Facts, Figures, and Future. *IEEE Computer,* 42(4), pp. 42-52.

Ebner, M., 2004. *TTCN-3 Test Case Generation from Message Sequence Charts.* Göttingen, Germany,: In Workshop on Integrated-reliability with Telecommunications and UML Languages (ISSRE04:WITUL}.

El Saddik, A., 2018. Digital Twins: The Convergence of Multimedia Technologies. *IEEE MultiMedia (Volume: 25 , Issue: 2 , Apr.-Jun. 2018),* 25(2), pp. 87 - 92.

Engeler, E., 1981. Algebras and Combinators. *Algebra Universalis,* pp. 389-392.

Engeler, E., 1995. *The Combinatory Programme.* Basel, Switzerland: Birkhäuser.

Engeler, E., 2019. Neural algebra on "how does the brain think?". *Theoretical Computer Science,* Volume 777, pp. 296-307.

ETSI European Telecoms Standards Institute, 2018. *TTCN-3 Standards.* [Online] Available at:
http://www.ttcn-3.org/index.php/downloads/standards
[Accessed 11 Dec 2018].

European Commission, 2010. *Directive 2010/30/EU of the European Parliament and of the Council of 19 May 2010 on the indication by labelling and standard product*

information of the consumption of energy and other resources by energy-related products. [Online]
Available at:
https://eur-lex.europa.eu/LexUriServ/LexUriServ.do?uri=OJ:L:2010:153:0013:0035:EN:PDF
[Accessed 11 Dec 2018].

Fathi, B., 2017. Towards a Methodology for Performance Evaluation in Terminology Planning. In: P. Faini, ed. *Terminological Approaches in the European Context.* Newcastle upon Tyne, UK: Cambridge Scholars Publishing, pp. 328-347.

Fehlmann, T. M., 1981. *Theorie und Anwendung der Kombinatorischen Logik,* Zürich, CH: ETH Dissertation 3140-01.

Fehlmann, T. M., 2003. *Linear Algebra for QFD Combinators.* Orlando, FL, International Council for QFD (ICQFD).

Fehlmann, T. M., 2016. *Managing Complexity - Uncover the Mysteries with Six Sigma Transfer Functions.* Berlin, Germany: Logos Press.

Fehlmann, T. M., 2019. *Cloud Samples for ART.* [Online]
Available at: https://web.tresorit.com/l#QtOlbhUCcQB-oVG7hSatSQ
[Accessed 04 08 2019].

Fehlmann, T. M. & Kranich, E., 2011. *Transfer Functions, Eigenvectors and QFD in Concert.* Stuttgart, Germany, QFD Institut Deutschland e.V.

Fehlmann, T. M. & Kranich, E., 2012. *Using Six Sigma Transfer Functions for Analysing Customer's Voice.* Glasgow, UK, Strathclyde Institute for Operations Management.

Fehlmann, T. M. & Kranich, E., 2014-1. *Exponentially Weighted Moving Average (EWMA) Prediction in the Software Development Process.* Rotterdam, NL, IWSM Mensura.

Fehlmann, T. M. & Kranich, E., 2014-2. *Uncovering Customer Needs from Net Promoter Scores.* Istanbul, Turkey, 20th International Symposium on Quality Function Deployment.

Fehlmann, T. M. & Kranich, E., 2017. *Autonomous Real-time Software & Systems Testing.* Göteborg, s.n.

Gigerenzer, G., 2007. *Gut Feelings. The Intelligence of the Unconscious..* New York, NY: Viking.

Graz University of Technology, 2018. *Meltdown and Spectre.* [Online]
Available at: https://meltdownattack.com
[Accessed 5 Nov 2019].

Greenberg, 2015. *Hackers Remotely Kill a Jeep on the Highway – With Me in It.* [Online]
Available at: https://youtu.be/MK0SrxBC1xs
[Accessed 15 March 2018].

Gunning, D., 2017. *Explainable Artificial Intelligence (XAI)*. [Online]
Available at: https://www.darpa.mil/program/explainable-artificial-intelligence
[Accessed 22 Mar. 2019].

Gunn, S., 1998. *Support Vector Machines for Classification and Regression,* Southampton: ISIS Technical Report, University of Southampton.

IFPUG Counting Practice Committee, 2010. *Function Point Counting Practices Manual - Version 4.3.1,* Princeton Junction, NJ: International Function Point User Group (IFPUG).

Ishikawa, K., 1990. *Introduction to Quality Control.* Translated by J. H. Loftus; distributed by Chapman & Hall, London ed. Tokyo, Japan: JUSE Press Ltd.

ISO 16355-1:2015, 2015. *ISO 16355-1:2015, 2015. Applications of Statistical and Related Methods to New Technology and Product Development Process - Part 1: General Principles and Perspectives of Quality Function Deployment (QFD),* Geneva, Switzerland: ISO TC 69/SC 8/WG 2 N 14, Geneva, Switzerland: ISO TC 69/SC 8/WG 2 N 14.

ISO 26262-1, 2011. *Road vehicles - Functional Safety - Part 1: Vocabulary,* Geneva: ISO/TC 22/SC3.

ISO 31000:2018, 2018. *Risk management − Guidelines,* Geneva, Switzerland: ISO/TC 262.

ISO/IEC 14143-1:2007, 2007. *Information technology - Software measurement - Functional size measurement - Part 1: Definition of concepts,* Geneva, Switzerland: ISO/IEC JTC 1/SC 7.

ISO/IEC 19761:2019, 2019. *Software engineering - COSMIC: a functional size measurement method,* Geneva, Switzerland: ISO/IEC JTC 1/SC 7.

ISO/IEC 20926:2009, 2009. *Software and systems engineering - Software measurement - IFPUG functional size measurement method,* Geneva, Switzerland: ISO/IEC JTC 1/SC 7.

ISO/IEC CD Guide 98-3, 2015. *Evaluation of measurement data - Part 3: Guide to uncertainty in measurement (GUM),* Geneva, Switzerland: TC/SC: ISO/TMBG.

ISO/IEC Guide 99:2007, 2007. *International vocabulary of metrology – Basic and general concepts and associated terms (VIM),* Geneva, Switzerland: TC/SC: ISO/TMBG.

ISO/IEC/IEEE 29119-4, 2015. *Software and systems engineering − Software testing − Part 4: Test techniques,* Geneva, Switzerland: ISO/IEC JTC 1.

ISTQB, 2011. *ISTQB - Certifying Software Testers Worldwide.* [Online]
Available at: http://www.istqb.org/downloads/category/2-foundation-level-documents.html
[Accessed 24 April 2017].

ISTQB, 2014. *Agile Tester Extension Syllabus.* [Online]
Available at: http://www.istqb.org/downloads/send/5-agile-tester-

extension-documents/41-agile-tester-extension-syllabus.html
[Accessed 24 April 2017].

Mazur, G., 2014. *QFD and the New Voice of Customer (VOC)*. Istanbul, Turkey, International Council for QFD (ICQFD), pp. 13-26.

Mazur, G. & Bylund, N., 2009. *Globalizing Gemba Visits for Multinationals*. Savannah, GA, USA, Transactions from the 21st Symposium on Quality Function Deployment.

Myers, R. H., Montgomery, D. C. & Anderson-Cook, C. V., 2009. *Response Surface Methodology: Process and Product Optimization Using Designed Experiments*. New York, NY: John Wiley & Sons.

Oriou, A. et al., 2014. *Manage the automotive embedded software development with automation of COSMIC*. Rotterdam, NESMA.

Poppendieck, M. & Poppendieck, T., 2007. *Implementing Lean Software Development*. New York, NY: Addison-Wesley.

Pupale, R., 2018. *Support Vector Machines (SVM) - An Overview*. [Online]
Available at: https://towardsdatascience.com/https-medium-com-pupalerushikesh-svm-f4b42800e989
[Accessed 28 Mar. 2019].

Reichert, A., 2015. *Testing APIs protects applications and reputations*. [Online]
Available at: https://searchsoftwarequality.techtarget.com/tip/Testing-APIs-protects-applications-and-reputations
[Accessed 4 Apr. 2019].

Reichheld, F., 2007. *The Ultimate Question: Driving Good Profits and True Growth*. Boston, MA: Harvard Business School Press.

Rouse, M., Burns, E. & Laskowski, N., 2018. *Essential Guide*. [Online]
Available at: https://searchenterpriseai.techtarget.com/definition/AI-Artificial-Intelligence
[Accessed 12. Sep. 2019].

Russo, L., 2004. *The Forgotten Revolution - How Science Was Born in 300 BC and Why It Had to Be Reborn*. Berlin Heidelberg New York: Springer-Verlag.

Saaty, T. L., 1990. *The Analytic Hierarchy Process – Planning, Priority Setting, Resource Allocation*. Pittsburgh, PA : RWS Publications.

Saaty, T. L., 2003. Decision-making with the AHP: Why is the principal eigenvector necessary?. *European Journal of Operational Research*, Volume 145, pp. 85-91.

Saaty, T. L. & Alexander, J. M., 1989. *Conflict Resolution: The Analytic Hierarchy Process*. New York, NY: Praeger, Santa Barbara, CA.

Schurr, S., 2011. *Evaluating AHP Questionnaire Feedback with Statistical Methods*. Stuttgart, Germany, 17th International QFD Symposium, ISQFD 2011.

Schwaber, K. & Beedle, M., 2002. *Agile Software Development with Scrum*. Upper Saddle River, NJ: Prentice Hall PTR.

SonarSource S.A, Switzerland - Open Source, 2017. *Documentation for SonarQube 6.3.* [Online]
Available at: https://docs.sonarqube.org/
[Accessed 21 April 2017].

Soubra, H., Abran, A. & Ramdane-Cherif, A., 2014. *Verifying the Accuracy of Automation Tools for the Measurement of Software with COSMIC – ISO 19761 including an AUTOSAR-based Example and a Case Study.* Rotterdam, s.n.

Soubra, H., Abran, A. & Sehit, M., 2015. Functional Size Measurement for Processor Load Estimation in AUTOSAR. *Lecture Notes in Business Information Processing,* Volume 230, pp. 114-129.

Staimer, M., 2015. *TechTarget Essential Guide.* [Online]
Available at: https://searchdatabackup.techtarget.com/tip/Docker-data-container-protection-methods-Pros-and-cons
[Accessed 15 Jan. 2019].

Steve Singh et.al., 2018. *Docker overview.* [Online]
Available at: https://docs.docker.com/engine/docker-overview/
[Accessed 9 April 2018].

Szegedy, C. et al., 2014. *Intriguing properties of neural networks.* [Online]
Available at: https://arxiv.org/abs/1312.6199
[Accessed 8 March 2018].

The Kubernetes Authors, 2018. *Kubernetes.* [Online]
Available at: https://kubernetes.io
[Accessed 15 Dec 2018].

The National Transportation Safety Board, 2018. *Preliminary Report Highway Hwy18mh010.* [Online]
Available at: https://www.documentcloud.org/documents/4483190-NTSBuber.html
[Accessed 13 Mar. 2019].

Theodorou, A., Wortham, R. & Bryson, J., 2017. Designing and implementing transparency for real time. *Connection Science,* 29(3), pp. 230-241.

Tilghman, C., Li, M. C. & Zemore, M., 2014. *Software Safety Analysis Procedures.* St. Louis, MO, International System Safety Training Symposium.

Turing, A., 1937. On computable numbers, with an application to the Entscheidungsproblem. *Proceedings of the London Mathematical Society,* 42(Series 2), p. pp 230–265.

van Gerven, M., 2017. *Computational Foundations of Natural Intelligence.* [Online]
Available at:
https://www.frontiersin.org/articles/10.3389/fncom.2017.00112/full
[Accessed 27 March 2018].

van Gerven, M. & Bothe, S. eds., 2018. *Artificial Neural Networks as Models of Neural Information Processing.* Lausanne, Frontiers Media.

VDA, 2008. *Teil FMEA Band 4,* Berlin: Verband der Automobilindustrie.

Volpi, L. & Team, 2007. *Matrix.xla.* [Online]
Available at: http://www.bowdoin.edu/~rdelevie/excellaneous/matrix.zip
[Accessed 11 Dec 2018].

Wikipedia, 2018. *Blockchain.* [Online]
Available at: https://en.wikipedia.org/wiki/Blockchain
[Accessed 17 April 2018].

Wikipedia, 2019. *European Train Control System.* [Online]
Available at:
https://en.wikipedia.org/wiki/European_Train_Control_System
[Accessed 13 Feb. 2019].

REFERENCE INDEX